Somerset Beekeepers and Beekeeping Associations

A History 1875–2005

David Charles

© A. D. Charles 2005

All rights reserved. No part of this publication may be reproduced, stored in a retrieval system, or transmitted in any form or by any means (electronic, mechanical, photocopy, recording or otherwise) without the prior permission in writing of the author.

Published by Somerset Beekeepers' Association

ISBN 0905652–66–5

Trade Distribution by BBNO

Printed by Direct Offset, 27c High Street, Glastonbury, Somerset BA6 9DR

Contents

                                                                                             Page

Preface		4
Acknowledgements		5
Foreword		6
Chapter 1:	Associations of short Duration	7
Chapter 2:	The present Association is inaugurated	25
Chapter 3:	New Rules and Advancement	35
Chapter 4:	The 1930s, but another World War follows	45
Chapter 5:	The early Post-war Years	51
Chapter 6:	The 1960s and 1970s	61
Chapter 7:	The 1980s: a Decade of Decline	73
Chapter 8:	Towards the Millennium	81
Chapter 9:	The Millennium and Beyond	91
Chapter 10:	Presidential Profiles	99
Chapter 11:	Some prominent Members	125
Chapter 12:	Miscellany	141
Appendix I:	Principal Officers	152
Appendix II:	Isle of Wight Disease	155
Sources of Reference		157
Index		158

Preface

It became evident to me quite early in my membership that the Somerset Beekeepers' Association possessed a very interesting past. While browsing through early post-war copies of the Association's year books I came across a series of articles published in three successive years, 1950, 1951 and 1952 outlining history, not only of this Association, but also of its predecessors of which there appeared to be three. These articles were researched and written by Mr A. R. Baker who was headmaster of Nunney School, near Frome, and a member of the Eastern Division. He wrote under the name of "Zummerzet." His account related the course of events over the first fifty-five years.

Looking back is considered an undesirable trait, but there are times when a yearning to learn more of what brought us to where we stand today becomes an irresistible temptation to probe. My immediate thought was that Baker's fascinating chronicle must be continued so that events are recorded up until the present. That was fifteen years ago in 1990, and left more years to be researched than had already been covered - a period of sixty years. My intention, with the Association's support was simply to continue from where Baker had ceased and append my work to his. My life over the next years was too full to give the matter much attention but during this time a file of references was gradually accumulating.

On commencing research in earnest I gradually realised that there was more of interest from those early days unrecorded in Baker's account. Although some of his text is still evident and forms the core of the early account, I have researched all those years again. In consequence much of Baker's text has been amended and substantial additions made. Baker deserves credit not just for those words which are his, but also for having been the catalyst sparking my enthusiasm to continue from the point where, having reached 1930, he put down his pen.

David Charles
West Pennard, Somerset
November 2005

Acknowledgements

Many friends and institutions have been of considerable help to me in producing this book. My close friend Marion Parsons has taken a keen interest throughout over two years of research and writing. Her constructive comments on the text and assistance with reading the proofs have been invaluable. I am indebted to Pat Rich of Chilcompton for writing the Foreword and a profile for Chapter 10. David and Ann Morris scoured several journals, photocopied relevant passages and allowed me to borrow photographs. Other people to whom I am indebted for material include Richard Bache, Kenneth Edwards, Gerald Fisher, David Hounsell, John Newcombe, Will Messenger and David Pearce, but there are many more people who have assisted in some small way.

It has been a privilege to have sight of old minute books of the British Beekeepers' Association and the South-West Counties Joint Consultative Council. I am grateful for the co-operation of the International Bee Research Association at Cardiff, the Director, Richard Jones and the staff there, also to the Somerset and the Northamptonshire Records Offices for access to documents and journals. Finally, I cannot overlook mentioning the printers, Bill and Rowena Wallace at Direct Offset, Glastonbury, for their advice, patience and skill in producing the book. My hope is that readers will feel this has been a worthwhile project.

David Charles
West Pennard, Somerset
November 2005

FOREWORD

The celebration of the Centenary of the Somerset Beekeepers' Association in 2006 presents a wonderful opportunity to reflect upon the past one hundred years of our history and of the impressive contribution to British beekeeping at both county and national level.

I am privileged to write a foreword to a book about this history, which I know by the nature of the author, David Charles, will have been meticulously researched, from material in diverse locations, much of it ephemeral, to produce a history rich in facts and devoid of conjecture. He will have travelled many miles and spent countless hours in pursuit of the true facts.

David Charles, a beekeeping friend for close on twenty years, gave me, our current President Mary Barnes-Gorell and many others, a sound early grounding in the craft of beekeeping and even now I still greatly appreciate the tremendous gift of encouragement, expertise and time which he gave to all. A Master Beekeeper and as top candidate, a holder of the Wax Chandlers award, I well remember his visits to inspect and advise on now seemingly simple problems in my own apiary, with no thought of personal reward except that he was helping a fellow new beekeeper along the way.

I have had an opportunity to read parts of the early draft of this history and now look forward in eager anticipation to the finished book. It is a unique insight into our past and had it not been written now, much of the included material would have been lost in the mists of time. It also comes in a period of great challenge to the present generation of beekeepers, whom it is hoped will be enlightened and inspired by the deeds of our forbears and will therefore confidently cope with the new threat which the varroa mite poses in our present time.

Congratulations to David on producing this history of Somerset beekeepers in such timely circumstances. With the knowledge of our illustrious past history, which this book portrays, and the outstanding contributions our predecessors have made, we are then able to look forward and to confidently anticipate our future in this fascinating craft of beekeeping.

Patrick Rich N.D.B.
Chilcompton, Somerset
November 2005

Chapter 1:
Associations of short Duration

The 1870s: the Scene is set, the Seeds are sown

Our story begins at a time when bees were generally kept in straw skeps and boxes, and only a few enlightened persons, mainly among the gentry and clergy, were using movable frame hives. Sons learnt from their fathers and were content to continue keeping bees in much the same way. Swarming was uncontrolled; some still harvested honey by killing the bees over a sulphur pit. There was no public instruction, little knowledge of bee diseases, and a general disregard of the value of bees as pollinators. For cottagers beekeeping was not generally a hobby, but was an essential part of the domestic economy.

Dorset skep and cap made locally by a gipsy in traditional pattern.   *Photographs courtesy of IBRA*

In May 1873 Mr Charles Nash Abbott commenced publication of the British Bee Journal and in the following year, during the Crystal Palace Exhibition of 1874, eight beekeepers met at 168 Camden Street, London, and formed themselves into a Society called the British Bee-keepers' Association (BBKA).

The aim of the newly formed Association was expressed in the following words: *"whose object should be the encouragement, improvement, and advancement of Bee-Culture in the United Kingdom, particularly as a means of bettering the condition of cottagers and the agricultural labouring classes, as well as the advocacy of humanity to that industrious labourer – the Honey Bee"*. In short, the task was to effect a general change from the use of the skep to the management of the bar frame hive.

Mr Charles Tite was the prime mover in the formation of a Somerset County Association. He was a member of the BBKA and in September 1875, after visiting an exhibition of bees, hives and honey at Grantham, Lincolnshire, he proposed the formation of *"County*

Associations of Bee-keepers". He felt that this was the most effective way of promulgating knowledge of the craft.

The British Bee Journal announced that Mr C. Tite of Yeovil and Mr W. N. Griffin of Alphington, Exeter, were willing to become secretaries of their respective counties. Later, in the British Bee Journal of November 1st 1875 it was stated that Mr O. Poole of Uphill, Weston-super-Mare had invited the beekeepers of Somerset, Dorset and Devon to unite in forming a West of England Apiarian Society. William Griffin, however, was confident that Devonshire would form an Association of its own and a set of rules was drawn up at a preliminary meeting held on 17th of this month. This being the case, Charles Tite, living in Yeovil at the time because of his business, was willing to work for Somerset or Dorset.

In the issue of the Journal for December 1st 1875 it states that O. Poole was working vigorously in the formation of an Association and had just sent in a report. The report is not published; neither is the Association referred to by name.

It must be assumed from the available documentation that Poole succeeded in briefly establishing a Western Apiarian Society for Somerset only. This is in no way associated with or to be confused with the Devon based Western Apiarian Society that existed between 1797 and 1809.

Skep covered with roughly made hackle.

Photo courtesy of IBRA

The British Bee Journal for May 1877 states that arrangements had been made by the West of England Apiarian Society, Taunton Branch, to hold their second annual exhibition in Vivary Park together with the Horticultural Society. *"Valuable prizes will be offered for honey, bee-hives & etc. and there will be an extensive show of bee furniture, foreign bees, bees at work and interesting manipulations".* Unfortunately, owing to the unfavourable honey season this exhibition was cancelled. At Weston-super-Mare, however, Mr Obed Poole won several prizes. Charles Tite won 2nd prize for honey in sectional supers and also a bronze medal and 1st prize for the best collection of natural objects illustrating the natural history and economy of the honeybee. Thereafter Mr Poole seems to have disappeared from the scene, never to be heard of again.

The British Bee Journal of December 1881, under the heading **"Somersetshire – a Secretary wanted"** published an impassioned appeal by Charles Tite for a new Hon. Secretary for the West of England Apiarian Society, a portion of which reads as follows: *"The bee-keepers of Somerset number several hundreds, scattered all over the county, with here and there a good group, living close enough to act in concert. Amongst them, moreover, are many enthusiasts, and not a few whose names are well known in connexion with apiculture. Notwithstanding all these advantages, they have at present no county association. The West of England Apiarian Society, which had its head-quarters in Weston-super-Mare, and which made such a show a few years since, has practically ceased to exist … The fact is, Mr. A. L. Perrett, of Queen's Villa, Weston, who acted as secretary, treasurer, and committee (and did all the work admirably, too) is about to leave the neighbourhood, and cannot therefore continue the direction of affairs. Who will come to the rescue? It is a fine field, and could be easily managed by any gentleman who has his heart in the work and a little time to spare … The Somersetshire gentry, too, subscribe liberally to anything calculated to raise the status of their poorer neighbours. In short, a good working committee and a zealous secretary would soon place bee-keeping in its proper position throughout the county. Surely the men of Somerset will not be content to remain without organization now that Dorset and Wilts, Devon and Cornwall, are in working order."*

The gentleman who eventually came forward was the Rev. Walter Hook of Porlock.

Killing the bees over a pit containing burning sulphur. *Sketch by Jack Fieldhouse*

A second Attempt survives for a little longer

Probably an organised Somerset Bee-keepers' Association did not fully materialise until 1883. The main evidence is to be found in Volume 10 of the British Bee Journal, as follows:

10th November, 1882 *"Persons desirous of joining the Association for this county are requested to communicate their names, addresses and the amount of their subscriptions to the Rev. Walter Hook, the Rectory, Porlock, Somerset."* A similar notice appeared in the issue for April 1883 except that it commenced *"it is proposed to form a Bee-keepers' Association".*

15th June, 1883, a brief report on page 63 states *"The Association made considerable progress during the time of the Bridgwater Exhibition. Owing to ill health the Rev. W. Hook has been compelled to resign. The Rev. C. G. Anderson of Otterhampton Rectory has been appointed Hon. Secretary."* Anderson remained in this post until the Association's demise in 1888. He was an energetic organiser and lecturer, frequently demonstrating bee handling at shows.

To ensure that instruction was given by competent persons, the British Bee-keepers' Association established a scheme of examinations for "experts". The first positive reference to beekeeping examinations appeared in the British Bee Journal of July 1882 reporting that an examination was to be held on the first Monday in August and that the applicants were to state *"age, experience and to submit satisfactory testimonials of character, sobriety, good temper, orderly conduct, cleanliness, industry and attention to detail".* The examination was to be both verbal and written, the candidates being required to drive bees, transfer them to movable comb hives, make artificial swarms and be skilled in hive construction.

These "experts" were categorised as first, second or third class. The first reference to an expert operating in Somerset is to be found in an article by John Morland in the Street Village Album, Volume 35, No.4 of 1883. It is entitled "Four Pages on Bees". In this article he refers to a lecture given in Street by the Rev. W. E. Burkitt, Hon. Secretary and expert of the Wiltshire Bee-keepers' Association. *"Mr Burkitt"*, he says, *"having the charge of a Wiltshire Parish of 180 souls naturally finds he has time to spare, and this year is touring in Somerset to press forward a Bee-keepers' Association."* In the British Bee Journal of October 15th 1885 there is an article by Burkitt himself as follows:

"In the spring of 1883 I was requested by the late Rev. H. R. Peel, (then Secretary to the BBKA and newly appointed editor of the British Bee Journal) *to hold (as he called it) a 'Bee Mission' in Somerset, with a view to the formation of a*

Rev. W. E. Burkitt

County Bee-keepers' Association. A provisional committee was formed, and it was arranged that I should give an evening lecture at five different centres, visiting and assisting bee-keepers during the day.

The last place was Wells where Mr. Peel met me. The Lord Bishop of the Diocese presided at my lecture in the town hall, and most kindly consented to become the first President of the Association in which he has ever since taken much interest.

Last year 1884 at the invitation of Mr. F. J. Clark the district Hon. Sec., I paid a second and most pleasant visit to Street near Glastonbury; and at the end of last month, (September 1885) he again invited me, bee work in Wiltshire having made it impracticable earlier in the season, or to spend more than three nights under his hospitable roof. I believe I am right in saying that all the members of the firm of C. & J. Clark & Co. are practical and advancing bee-keepers, and are doing their best to encourage those in their employ (and other neighbours also), in the pursuit, both by their own example, and by helping to procure from the best makers good hives and appliances for all who desire them. Mr F. J. Clark met me at the station on September 29, and immediately after lunch the visitation of bee-keepers commenced. Then a pleasant drive to Butleigh, where Mr. Clark had arranged for me to do some expert's work, and give a lecture in the schoolroom, at which the Rev. C.G. Anderson, County Hon. Secretary, attended to advocate the claims of the Somerset B.K.A. and some new members were enrolled. The next day expert work went on busily from 8.30 a.m. till 5 p.m.; then a beautiful drive through Glastonbury and Wells to Wookey Hole, to dine with J. Hodgkinson, J.P., who kindly took the chair at my lecture in the schoolroom.

On Thursday morning another visit was paid to Butleigh at the request of some who had attended the lecture on Tuesday, to put stocks in order for winter; but heavy showers much impeded the work. In the evening a third lecture was given at Street in the Friends' meeting-house, attended by most of the local bee-keepers, many of whom begged for my help before leaving the next day.

Bee-keeping has advanced most satisfactorily in Street and the other villages I visited on this occasion; and if in other districts Mr. Anderson has such as able and energetic assistant hon. sec. as Mr. F. Clark, he is much to be congratulated. Altogether I visited fourteen bee-keepers owning seventy-three stocks in bar-frame hives, and four in old-fashioned skeps. Many of the frame-hives were good ones by our best makers, but far too many too roughly made for satisfactory working.

I was rather sorry not to see any attempt at working sections on skeps by those who cannot afford frame-hives. Most of the stocks I examined were strong and well provided for winter.

Without exception, all I visited seemed anxious to gain all the hints I could give them, and also to improve in bee-culture. My visit was a most pleasant one, and I hope it may lead to greater care and neatness in the apiary, and result in many joining the Association, who as yet do not seem to realise the advantages of membership so much as could be desired.

Beekeeping in this district has increased and improved so much since the formation of the Association (only two and a half years ago), considering the little time that Mr. F. Clark can devote to it, that I feel bound to send this report to the B.B.J., for the encouragement of other district secretaries – W. E. BURKITT, Hon. Sec. and Expert of Wilts B.K.A., Buttermere Rectory, Hungerford, Oct.8, 1885"

Skep supered with sections as recommended by Rev. W. Burkitt for those who could not afford frame hives. *Photo: BBKA Jubilee Book*

Shows and Exhibitions promote the Craft

The membership of the Somerset BKA in 1884 was 82. During that summer there was a bee exhibition at Stogursey where the Rev. C. G. Anderson acted as expert. The Association possessed two tents. The first housed the honey show, bee appliances, etc. *"through which the public passed by a gangway to the new bee tent of the Somerset B.K.A."* This tent, in use for the first time, had been made for the Association by Messrs. John Waddon & Son, rope and sacking manufacturers, of Eastover, Bridgwater. It was made under the direction of the Hon. Secretary himself and *"is capable of holding one hundred spectators. The tent is an hexagon; the spectators are protected from the weather by a waterproof roof, and from the bees, by transparent netting, while the bees have free access to the open air. The tent is complete in every respect. One of the special points is the ease with which it can be erected and dismantled".*

In May 1885, the first Somerset BKA honey show was held in conjunction with the

Associations of short Duration

Bee tent. Illustration from "A Book about Bees", Rev. C. F. G. Jenkyns, 1886. The tent made for Somerset BKA by John Wadden & Son, Bridgwater, would have been similar to this one.

Somerset Agricultural Show at Taunton. The Rev. C. G. Anderson and Mr T. B. Blow demonstrated in the bee tent. In the British Bee Journal there is a report of the 1887 Taunton Flower Show that says *"altogether the Bee and Honey Department was most attractive, and the large tent devoted to it was crowded with spectators"*. There is also a report of the Somerset Cottage Garden Society's show at Wellington. The honey section was not very large here. Most of the exhibitors had made use of large bell glasses for their surplus honey *"But it was evident that local beekeepers have yet much to learn, and that there is a fine field open in the neighbourhood for a new branch of the Somerset Bee-keepers' Association when the Hon. Secretary can spare time to set it afloat."*

The following year, also in mid-May, the Somerset Agricultural Show was held at Wells *"prettily situated in the Bishop's park"* (the Palace grounds). In addition to the usual honey classes there were classes for hives and entries were received from as far afield as Kent. The hives of Mr A. F. Hutchings did not arrive in time for the judging through some unfortunate delay on the railway. The show was, of course, organised by the indefatigable Rev. C. G. Anderson and judged by Messrs Wm. N. Griffin of Freshford and F. Clark of Street.

There were classes for honey at the Somerton Horticultural and Poultry Show held on 18th August at Somerton Erleigh, then the country seat of Colonel Pinney. It is stated in British Bee Journal of 26th August, that *"what was exhibited showed the advance of modern ideas in this pursuit, as it was bottled in nice 1 lb or 2 lb. bottles neatly tied over with dry parchment, and nicely staged."* There were classes for amateurs and cottagers and they included ones for a super from a straw skep, comb honey (one dozen sections), extracted honey (one dozen 1 lb bottles or six 2 lb bottles) and, believe it or not, for a collection of queen wasps, not less than one hundred, and, for those who could not manage this high figure, not less than fifty.

Also in 1886 Anderson published a fifteen page paper-covered book, illustrated by quaint line drawings, explaining the beekeeping appliances and terms of those days. Directed towards children and the less literate it was entitled "The Beekeepers' Alphabet". This book is very rare.

The Association's Demise

The Annual Meeting of 1888 was held at Bath on 26th January. Only one member besides the Rev. C. G. Anderson, Hon. Secretary, put in an appearance and no business was transacted. The position of the Association was then as follows: members of the Association, 39; Taunton Branch, 28: Bedminster Branch, 10; total, 77. Receipts were £24. 18s. 6d. (£24.92½), expenditure was £36. 2s. 11d. (£36.14½), leaving a deficit of £11. 4s. 5d. (£11.22). The work of the Association had been crippled both from want of funds and energy among the members. Shows had been held at Bath, Bedminster, Taunton, Dunster, North Petherton, Stoke Courcy, Nunney &c. The Rev. Charles G. Anderson, having resigned the Hon. Secretary and Treasurerships, was no longer responsible for the Association in any way. No business being transacted at the meeting, both offices became vacant. No plan for carrying on the Association was proposed, though a letter was read from Mr Hamilton Palairet, offering to assist in paying off the debt due to the Treasurer.

Thus the parent Association endured for about five years, and became inactive. Some of its northern members drifted into the Bristol Association but there is no doubt that it never really lost its entity as it existed among small gatherings and associations of beekeepers in various parts of Somerset and in liaison with other counties.

The Taunton Group continues to flourish

The Taunton Branch was a keen and active group. At the annual meeting held on February 16th 1888 the Chairman reported increased success and progress during the past year. Arrangements had been made with the parent body for the management by the branch of its own affairs subject to approval of the County Hon. Secretary and the payment of an affiliation fee fixed at 25% of the five shillings (25p) subscription. When the annual report of the County Association was read, general regret and surprise was expressed that Somerset beekeepers had allowed the Association to reach such a low ebb, notwithstanding the strenuous efforts of their zealous Hon. Secretary, the Rev. C. G. Anderson.

Associations of short Duration

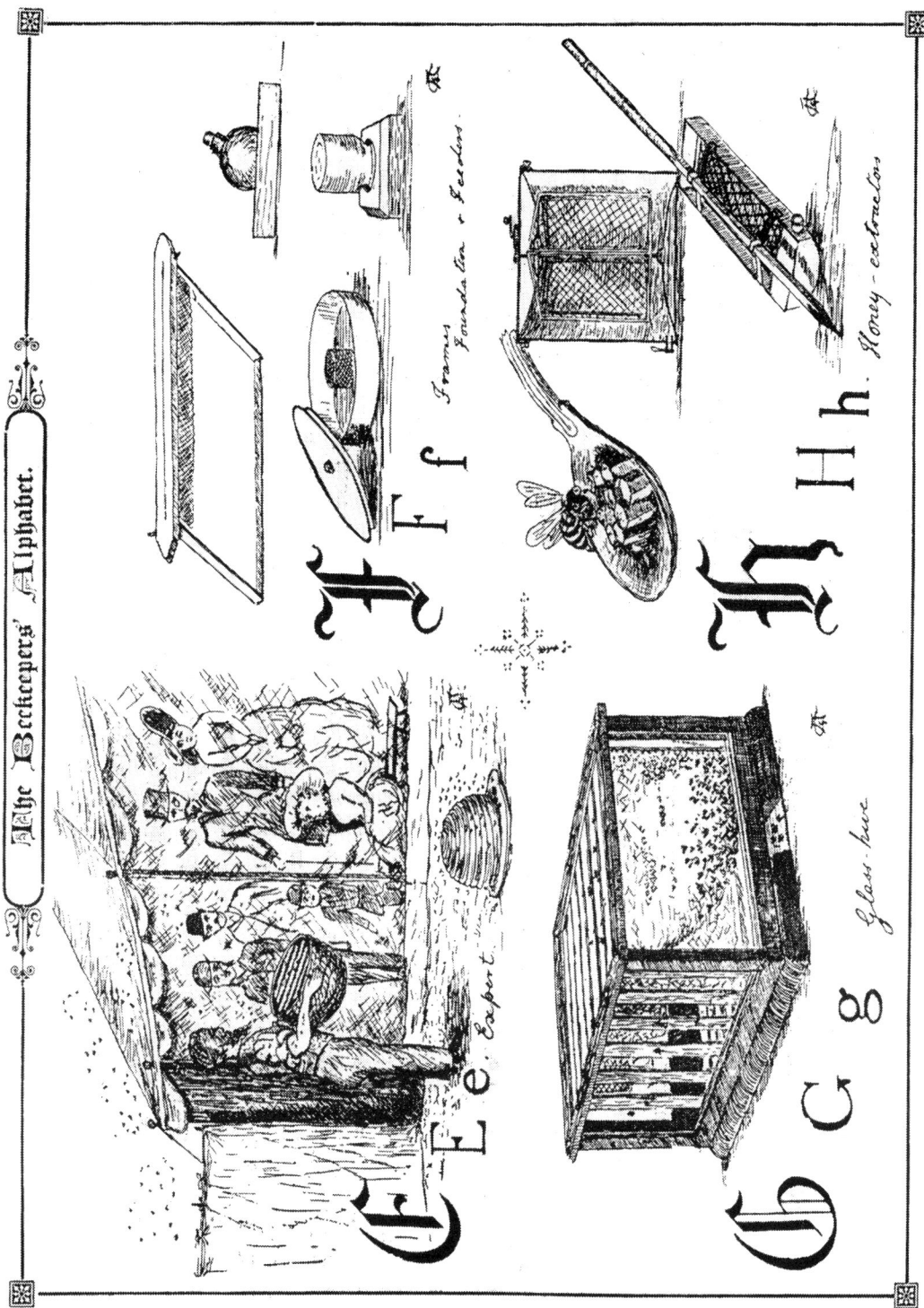

A page from Anderson's *The Beekeepers' Alphabet*

Apiary showing transition to wooden hive. *Photo courtesy of IBRA*

It was felt that an effort should be made to carry on the County organisation. A resolution was passed to this effect and another that *"Mr Anderson be invited to hold a general meeting of the County Association at Taunton on an early date"*. Alas, this action came too late. After the business Mr Tite delivered an address on "Bee-keeping for Pleasure and Profit" and referred to a survey done by the Rev. C. G. Anderson in 1884. Figures from various parts of the county showed that 230 stocks had produced an average yield of 57lb. He also seized the opportunity of publicly praising the self-denying labours of Mr Anderson and urged his listeners to do their utmost to spread a knowledge of humane bee-keeping amongst their friends and neighbours so that the use of superless straw skeps and the cruel custom of suffocating bees may soon become things of the past.

The group continued to be involved in organising the beekeeping section of the Taunton Deane Horticultural and Floricultural Society's show held at Vivary Park. At the twenty-second show, held on 15th August 1889 the beekeeping section was *"shorn of the attraction of manipulation"* but despite this it proved popular. There were several observatory hives and an excellent display of honey. The prize for the best collection of comb and run honey went to the well-known beemaster from North Petherton, Mr W. Pierce and Mr H. Barter of Taunton was second. Mr W. Withycombe of Bridgwater took a prize for his sections.

During the afternoon lectures on bees and beekeeping were delivered in the tent by the Rev. C. G. Anderson and Mr C. Tite.

In 1894 a special meeting was called to discuss the wording of a honey label to be used only by Taunton group members. The British Bee Journal also mentions that Rev. S. F. Cumming and Mr E. Chapman were to act as Hon. Secretaries for the Somerton and Taunton districts respectively.

As for the Rev. C. G. Anderson, the last seen reference to him was when he was asked to judge the Bristol District Bee-keepers' Association Show in July 1894. He died while still the incumbent at Otterhampton on 12th March 1898, aged 59.

The Bristol, Somersetshire and South Gloucestershire Association

The following extracts are taken from the original and authentic records of "The Bristol, Somersetshire and South Gloucestershire Bee-keepers' Association". This is the Association from which the present Somerset Beekeepers' Association evolved.

Before the formation of the Bristol, Somersetshire and Gloucestershire Bee-keepers' Association in 1889 there was an association formed in 1885 known as the Bristol District of the Gloucestershire Bee-keepers' Association. The start was a very small one and the growth at first slow, *"there being very few people in the neighbourhood who really knew what was happening inside their hives"*. County associations had been at work in many parts of the country spreading knowledge of beekeeping but at the time no county or other society had undertaken the work of reform among the apiaries of South Gloucestershire.

There had been village clubs which had done some good work, the modern bar frame hive being introduced in some cases. Generally speaking, however, the owners never availed themselves of the real uses of the movable comb hive, *viz.* to open them, remove or exchange combs to extract surplus honey, to lessen or increase living space for winter or summer, and to examine the condition of health, quantity of stores and number of bees. They managed their bar framed hives in much the same way as they did their straw skeps, except that occasionally a super was put on and removed at the end of the summer.

The apiarian of those days viewed the advantages of a beekeepers' society with a sceptical eye and lent an unbelieving ear to the explanations of the pioneers of this association but a few converts were obtained and eleven members united themselves into a society. With this small beginning a honey show could not be arranged but the bee tent was brought to the Frenchay Horticultural Show in 1885 held at Iron Acton in Captain Liddon's grounds when an able lecturer and expert, the Rev. E. Davenport, attended.

This gave an impetus to the society and the membership increased to 50 in the following year when nearly 600lb of honey were exhibited. After a few years of work under the County

Association it was found that the expense and loss of time in attending committee meetings at Gloucester caused the interests of the small local branch to be somewhat neglected and the advantage of taking a portion of North Somerset with Bristol as a central meeting place was so apparent that the Bristol, Somersetshire and South Gloucestershire Bee-keepers' Association was formed in 1889 with 110 members.

There was no doubt that Mr James Brown of Court Farm, Failand, and later of Bridge Street, Bristol, was the great moving force behind these arrangements. It was evidently the primary consideration to provide an association for Bristol, but the enthusiasm and ability of those who started it gradually influenced the surrounding areas and in ten years it had spread itself over a large portion of Northern Somerset.

Mr H. M. Appleton, the Hon. Secretary and a former committee member of the recently disbanded Gloucestershire BKA, stated at an Annual Meeting that *"the reason why this association had been started was because the Gloucestershire Association gave them very little assistance and the Somerset one had fallen through. They had therefore started an association which aimed to take in South Gloucester and North Somerset."*

It was not made clear in the records who actually made the first move, but the names of the following appear as the original officers: Lady Smyth (President), J. B. Butler (Chairman and Treasurer), H.M. Appleton (Secretary), J. Martin (expert), with a committee of H. W. Atchley, T. Butler, J. Collins, B. Hall, M. Harding and W. Fleming, together with the local secretaries for the following districts: Rev. Hugh Falloon (Ashton), W. G. Wyatt (Bedminster), H. H. Hamilton (Horfield), T. Hinton (Doynton), Mr Dixon (Clifton), W. E. B. Webley (Henbury), Mr Rawbone (Kingswood), Mr Rogerson (Bath) and G. Jarman (Knowle).

The first Annual General Meeting was held at Short's Coffee Tavern, 11 and 12 High Street, Bristol, on Thursday, April 17th, 1890, when J. B. Butler was in the chair.

A very businesslike system was evident from the outset. Demonstrations, lectures and honey shows took place in the Association's own tent at many of the large horticultural and flower shows. It is worthy of note that *"arrangements are made whereby the public can view with perfect safety the mysteries of the hive and witness the perfect command the scientific apiarian has over his bees"*. The name of the Rev. E. Davenport appears as a judge of honey. These events were sometimes preceded by music from a military band (The Bristol Rifle Volunteers) or a parish band and sometimes ended in a cricket match, e.g. Yatton versus Long Ashton on August 3rd, 1891.

The chief educational work was *"to impress and prove to the old "sceptists"* (sic) *how cruel was the practice of killing the bees to take their honey, and that it was quite possible to drive the bees into an empty hive which being provided with artificial comb allowed the bees to set to work at once and collect another store for the winter"*.

At the first annual dinner held at "The George & Railway Hotel," Victoria Street, Bristol, on February 17th, 1892, the following report was given:

"The number of members in 1890 was 110, and of this number six had left the district and nine had resigned which might possibly be accounted for on the ground that having derived knowledge from the association they no longer required its aid (laughter and hear hear*). On the other hand 75 new members had joined and the total membership was now 170. Anderson's bee tent had been in great demand, as many as three applications having been made for it on the same day. The tent had been erected and lectures given in practical beekeeping at the following shows – Totterdown and Knowle, Long Ashton, Keynsham, Kingswood and also at the North Somerset Agricultural Association's Show. Subscriptions and donations amounted to £36 and the total expenditure was £43.8s.11d (£43.44½)."*

A demonstration of bee-driving.   Photo: BBKA Jubilee Book

At the second Annual General Meeting Mr James (Jimmy) Brown of Baldwin Street (later High Street), became Hon. Secretary of the Association. There was a balance of £2. 10s. 0d. (£2.50). £2. 2s. 0d. (£2.10) was oversubscribed to the Bath and West Show and it was resolved that the Secretary and Committee should attend *"for the purpose of advertising our association"*. Honey shows formed a great feature in the beekeeping of those days and "Six or twelve bottles" was the order of the day and *not* "Two squat jars." Evidently these early exhibitors had to earn their prize money, but the benches were far more impressive than nowadays. In 1892 there were 170 members and over 800 hives were examined by one expert, Mr John Martin, during the season. He was a First Class Expert who shortly afterwards obtained the post of Bee Expert to the Government of South Africa.

The Technical Education Act of 1889 opened the door to county associations to obtain grants for their educational work from the new county and district education authorities. Following representations from the BBKA in 1891, beekeeping was included alongside agriculture and horticulture as a discretionary subject in suitable areas. The fact that this Association embraced not only two counties but also the city of Bristol was a problem in this respect. At a meeting on December 1st 1892, Mr Brown proposed *"that it may be advisable to change the name of the Association to the Somerset B.K.A., the object being to get a grant of money from the County Council for the purpose of giving lectures throughout the county."* For some reason this was not carried.

Also at this meeting, responding to an appeal in the British Bee Journal, it was decided to send some honey to the Chicago Exhibition. J. B. Butler and J. Brown both promised 5lb and this was sent to London for onward transit with the rest, the total amount sent from British beekeepers being nearly 1,000 lb. This refers to the occasion when the BBKA arranged to exhibit honey at the Chicago Exhibition. The undertaking was managed in London by a small committee chaired by Mr Cowan. The exhibit was well staged in a prominent position and received an award of merit.

In 1893 the Secretary in his report urged that *"it should be made clear to members that experts went round more to teach the members how to handle the bees than to do the menial work and carry out all the active arrangements, as some of them seem to think."*

In 1894 the Rev. J. Polehampton was elected as the local Hon. Secretary for the Frome District and Mr A. E. Martin as the expert.

By 1897 there were ten districts and there is no doubt that this association's work was spreading very rapidly in Somerset. As a result another expert was required, and an advertisement was placed in the British Bee Journal. Replies were received from many well known beekeepers, e.g. Mr Herrod (Newark) and Dr Bellew (Wellingborough). Mr W. Withycombe, Bridgwater, a man to be prominent in Somerset for the next fifty years, was appointed. He made 401 visits during that year and it was evident that this association was becoming a *"county affair"* for the Somerset ties were now the greatest. However much enthusiasm and interest was given to beekeeping, it would seem that foul brood disease was a source of continual worry in those days. Many beekeepers now imagine that those days of beekeeping were free from disease. The following are authentic extracts from the records:

March 2nd 1893, *"it was proposed that naphthaline should be supplied to each Expert free of cost for the purpose of suppressing foul brood."*

"With regard to the Improvement and Advance in Bee Culture the Experts' reports show that the movable frame system of bee-keeping is much more general. The wintering and feeding of bees

Associations of short Duration

George and Elizabeth Gifford, Box Cottage, Sticklynch, near West Pennard, circa 1890, still kept their bees the traditional way.
Photo courtesy of Mrs J. Croker

has been more satisfactory and better methods generally are being adopted, but the Association deplores the fact that in spite of preventative measures such as the free use of naphthaline, etc., foul brood is devastating many of our members' apiaries."

One member wrote *"unless we get legislation shortly it will stop beekeeping in this parish as every hive is more or less affected. Can we get legislative aid for the purpose of dealing with foul brood?"* (Report 1896). – Expenditure on naphthaline and carbolic was 11/6d (57½p) for the year.

At a meeting in February 1898 Mr James Brown showed some modern appliances and introduced samples of the Weed foundation and drawn comb as produced in America but which had not yet been put on the British market. *"When they do come they will work a revolution in beekeeping."* On 4th June, by invitation of the Hon. Secretary, Miss Hill-Dawe, a field day was held at "The Home", Long Ashton between two and eight o'clock. *"Interest in the proceedings was greatly enhanced by the presence of Lady Smyth, who, as President, took the chair at the afternoon meeting, though she had only returned the day before from her winter stay in India."* The bee tent was erected in the grounds, and in this some of the practical work of the apiary was illustrated. Mr Withycombe then gave an account of his spring visits. This year which opened so promisingly proved a disastrous one. In every part of

England, without exception, the occurrence of honeydew spoilt the honey harvest. What was gathered was unfit for sale and the year became known as *"the black season"*. Mr Langley of Radstock made a complaint that no expert had visited his district and it was decided that if he could guarantee twelve members, the Council would make it a new district.

The dawn of the Twentieth Century saw a very determined company of Bristol and Somerset beekeepers steadily forging ahead and spreading their influence far and wide. The success of their continued existence was plainly their intention to keep going at all costs. Not always was it plain sailing, for the indifference of some members caused trouble, whilst financial difficulties at times loomed very large. During 1899 for instance it was reported *"that a balance of £29. 7s. 11d (£29.39½) was due to the Treasurer, which was a very serious deficit and this lack of funds crippled the work of the Association … and although they received £22. 11s. 6d (£22.57½) special donations they ended up by being further in debt than ever."*

This was caused by the minimum subscription of one shilling (5p) for which the member received two visits from the experts. The committee had two courses open to them: either to raise the minimum subscription to two shillings (10p) or half a crown (12½p), or else allow only one expert visit. It was said the latter policy would prove disastrous, but the real trouble seemed *"that the members receive a spring visit from the expert and then wait and expect a second, whilst the Treasurer has to wait and wait and expect the member's subscription."*

There is no doubt however, that those in charge of the Association at that time were fine beekeepers and influential people with a very determined policy. In one case they did not consider a local expert *"competent or trustworthy"*, so they told him so and countermanded his appointment. In contrast, Mr Withycombe was doing splendid work over a large area.

Foul brood disease was very common and caused much devastation. Many remedies were tried, and Mr Jordan in an able manner at a tea and conversazione told his audience how to prevent and cure foul brood. Another expert stated he had burnt all his stocks on finding that they had the disease. However, many cases were always recorded in annual reports and the cure does not seem to have worked!

At the Annual Meeting held on January 29th 1900, it was reported that Miss Hill Dawe who had filled the office of Honorary Secretary and Treasurer since 1895 had resigned. She had done an excellent job of work and the Association was greatly indebted to her. Mr W. T. Tarr became Honorary Secretary *pro tem*.

During the year the experts' work had been undertaken by Messrs. Gough, Kirby, Sams, Willcox, Chapman and Withycombe. 1,363 stocks had been examined represented by 228 skeps, 1,101 top bar framed hives and 24 others. Foul brood cases numbered 79 and in some cases whole apiaries were affected.

The bee tent had been in attendance at shows at Knowle, Neston Park and Bradford-on-Avon. Receipts amounted to £84, expenditure £53 and thus the deficit of £30 adverse balance had been wiped out. The membership was now 216. The growth of the Association continued and required more detailed administration. A minute reads *"The Council finds it impossible to work the Bridgwater and Ilminster Districts without the aid of a local Honorary Secretary."* Honey shows were held at such places as Wells, Radstock, Chew Magna, Clutton, Bath and Knowle.

On April 26th, 1901 the books and cash were unexpectedly placed in the Council's hands, and Mr James Brown again became Honorary Secretary.

During 1902 the Secretary was instructed to offer two guineas (£2.10) to Mr W. Herrod to judge and lecture at the Clutton Show. During this year Mr J. Brown was taken seriously ill with enteric fever. A field day was again held at "The Home", Long Ashton, on August 2nd, 1902, which proved a great success for about seventy members were present to witness the bee driving.

The Annual Meeting was held at the Occidental Café, Bristol, on February 10th, 1903 with Mr Samuel Jordan as Chairman. Mr James Brown reported *"The Secretary regrets to report a large and growing class amongst beekeepers whose horizon is bounded by their garden wall and with short memories ... even forget to pay for the privileges they have freely enjoyed at the expense of the expert."* One member desired the Association to provide facilities for the disposal of honey amongst grocers. It was stated that *"8/6d (42½p) could be obtained for one dozen 1-lb tie over jars ... what was really wanted was an attractively packed 1-lb bottle with a neat label."*

The following were the named districts with the respective experts in 1903:

1. Bristol City, Abbots Leigh, Portishead, Bedminster, Long Ashton, Patchway, Chew Magna: G. Kirby.
2. Keynsham, Kingswood: W. P. Morris.
3. Yate: G. Wollen.
4. Bridgwater and Ilminster: W. Withycombe.
5. Bath: J. W. Brewer, First Class Expert and Headmaster of Walcot Council School.
6. Congresbury, Clevedon, Weston-super-Mare, Wells, Radstock, Frome: E. S. A. Gough.

The districts extended well beyond the places named, and thus may be seen how the Association was spreading into Somerset.

On May 29th, 1903, a Honey Show was held on Durdham Down where the large bee tent with its red flag flying was very evident. Mr W. Herrod, the well-known apiarist and expert lecturer of the BBKA was present to lecture and give demonstrations.

In 1905, Mr James Brown, owing to the pressure of his business was compelled to give up the office of Honorary Secretary. It was at this point that the name of Mr L. E. Snelgrove appears – a name that was destined to make such a mark and resounding influence on the future history of this and the future Somerset BKA. Mr Snelgrove had been associated with this work since 1902. He at first declined the secretaryship, at least for a little time and he was deputed to ask Mr Charlton to become secretary. However, on July 8th, 1905, at the age of 26, Mr Snelgrove was appointed. The balance of 15/4d (76½p) was all that remained for the new secretary to work upon and of this 10/6d (52½p) was a donation to help the Foul Brood Bill.

Mr Snelgrove was already a member of the Council and the visiting expert for Clevedon, Weston-super-Mare, Portishead and Wells. There is no doubt that when he took over as Hon. Secretary of the Bristol based Association in 1905 his great ambition was to lay the foundations of one huge Somerset Association, for a very tangled skein of various beekeeping influences existed at that time. Affiliated branches extended no further south than Bridgwater and Wells.

Mr Icingbell's apiary at Cheddon Fitzpaine. *Photo: B.B.J.*

Chapter 2:
The present Association is inaugurated

A Period of rapid Expansion, then War and Disease intervene

Mr Snelgrove's aim was the welding of all the various influences into one Somerset Association as proposed by Mr Brown some fourteen years previously. In 1906 this ambition became a reality, for the Somerset BKA was given another start or re-birth. At the 17th Annual Meeting held on February 13th it was resolved to reconstruct the Bristol District BKA as the Somerset BKA for it was the strongest organised body that could develop and aspire to be so. The inaugural meeting was held on May 3rd. The President and officers of the old Association became the officers of the Somerset Bee-keepers' Association with the exception of Mr Jordan who was succeeded as Chairman by Lt. Col. H. F. Jolly.

The objectives of the newly formed Association were *"the encouragement, improvement and advancement of bee culture, and incidentally of fruit culture, particularly as a means of bettering the condition of artisans and cottagers, as well as the advocacy of humanity to that most industrious of labourers – the honey bee."* The minimum subscription was set at five shillings (25p) per annum, and of *bona-fide* cottagers owning not more than six hives, two shillings and sixpence (12½p) per annum, payable in advance. Donors of £5 upwards became life members. Members desiring no visits from experts or for whom no such visits could be provided could enjoy all the other privileges of the Association on payment of 2/6d (12½p), and cottagers 1/- (5p) per annum.

Samuel Jordan of Bristol and J. W. Brewer of Bath worked very hard with Mr Snelgrove to re-establish the Association. These people were "First Class Experts" and they were very few and far between in those days. In 1906 the County Council Agricultural Committee consented to pay for evening lectures and demonstrations. Messrs Snelgrove, Jordan and Brewer were appointed to deliver these. Mr Withycombe was working the Bridgwater district, whilst Mr Kirby had the Kingswood, Patchway and Yate districts, and Mr Barwell visited the Abbots Leigh area. It should also be mentioned that Mr J. H. Burton, the County Agricultural Secretary, was very keen and helpful in this work of re-establishing the Somerset BKA. Thus Bristol then became a Branch and that is why the Bristol beekeepers were in the Somerset BKA until the Avon County Beekeepers' Association was formed in 1974.

A Period of rapid Growth

In 1906 the members numbered 75 but in 1908 the number was not far from 200. Under Mr Snelgrove's influence and ability a fresh enthusiasm prevailed and new branches and enthusiastic beekeepers appeared in all parts of the county. In 1907 the Frome and Radstock

John Spiller at his Taunton apiary, 1910.

Branches were formed with Mr P. B. Rigg and Mr H. J. Moore as respective secretaries. It would indeed be a long story to record the formation and progress of all the branches.

Portishead formed a branch in 1907 with Mr W. J. Lang as Hon. Secretary. In 1908 Shepton Mallet established a large branch of 30 members with Mr H. Grist as Hon. Secretary. Mr B. Hutchings as expert and local Hon. Secretary established what promised to be an even larger branch at Yeovil. Glastonbury and Street also formed a branch in 1908 with Mr R. Sims as expert and Mr H. B. Clark as Hon. Secretary.

Mr A. Hole of Wedmore established a branch covering the Wedmore, Wells and Cheddar districts. Wells formed its own branch under Mr Bigg-Wither a year later. Wrington and Castle Cary branches were formed in 1909 and Chard had its branch in preparation. In 1911 Mr Chick, aided by Messrs Cowan, Maynard and Tite received credit for building the Taunton Branch into the largest in the Association (49 members). Mr J. Spiller then took over from him as Hon. Secretary. In 1908 the Association had eleven branches, in 1909 sixteen, 1910 twenty and in 1911 there were twenty-three in total.

Thus it may be seen how great was the work in moving forward – and one name was the mainspring – Mr L. E. Snelgrove. This was recorded by the Somerset BKA. in 1909: *"The Council desires to place on record its highest appreciation of the able and laborious discharge of the duties of the Honorary Secretary, Mr. L. E. Snelgrove."* The Somerset Bee-keepers' Association was now holding shows, lectures and demonstrations even in little known places in the county and many famous and influential people were associated with its activities. Snelgrove must have been thrilled by this rapid rate of expansion and the county-wide enthusiasm he and colleagues had engendered. His role now took a change of direction.

It was in 1912 that Mr L. E. Snelgrove *"owing to a continuous increase in his professional duties and the concurrent increase in the responsibilities connected with the Association was compelled to resign his post as Honorary Secretary of the Somerset BKA"*. He had the satisfaction of seeing it multiply tenfold for it now had over 500 members. The financial state and organisation were both excellent and the Somerset BKA had every prospect of becoming the most influential beekeeping association in the country.

As a token of their appreciation the members presented Mr Snelgrove with a silver rose bowl costing £25. It is doubtful if the true measure of his influence, ability and work can ever be fully recorded and as long as the Somerset BKA endures, the name of L. E. Snelgrove will be remembered. Mr Lovelace Bigg-Wither of Wells was appointed as Honorary Secretary to succeed him.

The Somerset Bee-keepers' Association had the honour in its checkered history of having many famous names associated with its work. During this period, Mr T. W. Cowan, F.L.S., F.R.M.S., Chairman and later President of the British Bee-keepers' Association took a great interest in the Somerset BKA. Lady Smyth who had been the first President of the society formed in Bristol in 1889 continued as President until her death in November 1914. Following her death, Mr Cowan was elected President of the Somerset BKA on February 27th, 1915 and he held this office until his death in 1926.

The Great War Years and Isle of Wight Disease

The Annual General Meeting in 1913 was held at The Railway Hotel, Yatton, on February 8th, when Mr Cowan presided. It was reported that 1912 had been a bad honey season. The membership was *"round about 450."* Naphthalene was still supplied for the prevention of foul brood. It was at this meeting the Hon. Sec., Mr Bigg-Wither reported *"the dread 'Isle of Wight' disease had become more and more prevalent and that the Association must view it with grave concern"*.

It would appear that the Isle of Wight disease made its appearance in the Isle of Wight about the year 1904, reached the mainland 1907 and in a few years had spread to almost every corner of the country and swept beekeeping almost out of existence. Probably the true incidence will never be known for beekeepers just gave up and said no more about it. It was in 1911 that the disease was first noticed in Martock but in whose apiary is not stated. In 1912 the disease had appeared in several districts but in most cases the experts were given a free hand and each affected stock was destroyed, the hives and surrounding ground being

The Somerset Bee-keepers' Association.

(In Affiliation with the British Bee-keepers' Association.)

SHOW OF HONEY, BEES, Etc.,

IN

THE PARK, TAUNTON,

IN CONNECTION WITH THE

TAUNTON HORTICULTURAL & FLORICULTURAL SOCIETY,

ON

THURSDAY, AUGUST 1st, 1912.

Judges: T. W. COWAN, Esq., F.L.S., F.G.S., and S. JORDAN, Esq.

OPEN CLASSES.

First Prize given by Lady Smyth, Ashton Court:

179 For the finest Collection of Honey and Wax of any year, staged in the most attractive form on a space 3-ft. by 3ft. and height not to exceed 4-ft. above the table 20/- 12/6 7/6 5/-

Prizes given by the Rt. Hon. The Earl Waldegrave:

180 For the best 12 1-lb. Bottles of Extracted Honey 10/- 7/- 5/- 2/6

Prizes given by W. H. Bateman Hope, Esq.:

181 For the best 12 1-lb. Sections of Comb Honey 10/- 7/- 5/- 2/6

182*For the best 1-lb. Bottle of Honey (not granulated) ... 7/6 5/- 3/- 1/-

183*For the best 1-lb. Section of Comb Honey 7/6 5/- 3/- 1/-

184 For the best Collection of Beehives and Appliances to be staged by the Exhibitor or his representative on a table 12-ft. long and 2-ft. wide 15/- 10/- 5/-

185 For the best Exhibit of an Educational and Scientific Nature relating to Bee-keeping ... Prize not exceeding 10/-

Cottagers residing in Somerset and others who have previously exhibted at the Taunton Deane Flower Show as Cottagers, may exhibit in the above Open Classes, without Entry Fees, on complying with the conditions respecting Cottagers in the Schedule of the Flower Show which can be obtained from the Show Secretary.

Classes 186, 187 and 188 are for such Cottagers only, and are free of Entry Fees.

*No Entry Fee will be charged in Classes 182 and 183, but every Exhibit, whether winning a prize or not, will become the property of the Association, and will be offered for Sale.

Elworthy & Son, Bristol.

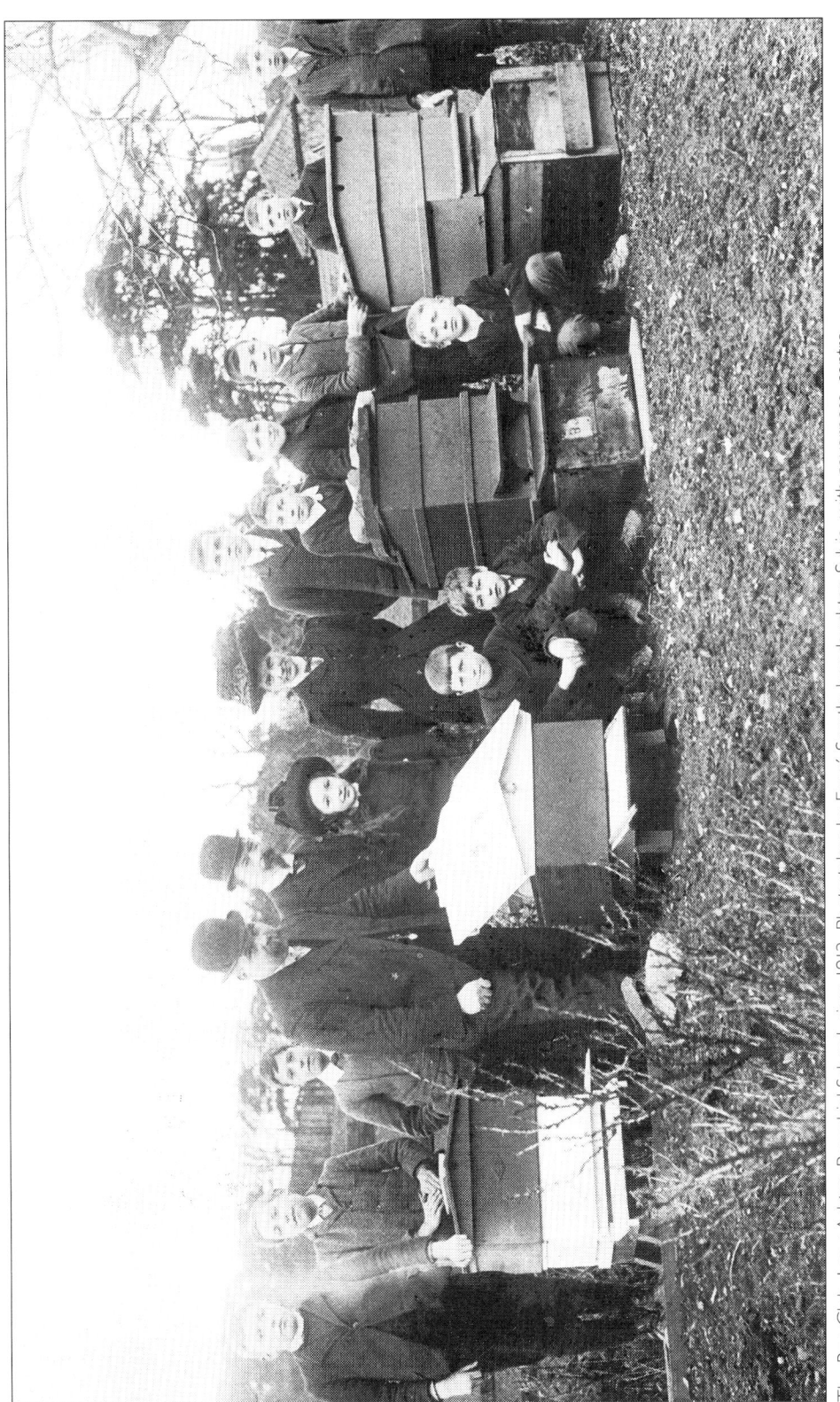

The Bee Club, Long Ashton Parochial School, circa 1912. Photo taken by Esmé Smyth; her daughter, Sylvia, with governess, centre.

Photo courtesy of the Malago Society, Bristol

thoroughly disinfected. Then it spread over the whole country *"leaving a trail of empty hives in its wake and many bewildered bee-keepers"*.

The disease was described at an Annual Meeting as *"a mysterious malady … evil days were these for the beekeepers … the bees came to the entrances with hooked wings and swollen bodies then fell off the alighting boards never to return … they crawled into large groups or clusters and there died. Beekeepers were powerless to stop the plague"* – and then this was added … *"the Government would no nothing".* In fact, the Board of Agriculture had established a scientific committee. Reports were published by HM Stationery Office in 1912 and 1913. The problem was under constant investigation and review. Legislation was under consideration and later there was the Government scheme for re-stocking from abroad in which Somerset participated.

This disease practically extinguished both the bees and the Somerset BKA. For some years from this date the story deals largely with Isle of Wight disease, and how it affected beekeeping in Somerset, together with the measures used to overcome the problem and the consequent re-stocking scheme for the county. Throughout it must be realised that the work of the Somerset BKA went on. There was a dogged perseverance of those in office to keep going and there is no doubt that these endeavours saved the Association from extinction during this ordeal.

We should also remember that the Great War of 1914–1918 occurred during this period and many beekeepers were serving in the armed forces and away from their homes, some for a considerable time, and others, never to return. The British Bee Journal regularly featured a Roll of Honour publishing details of any beekeeper serving his King and Country at home or abroad. The following names of Somerset beekeepers appeared on the dates shown. There were doubtless very many more.

Oct. 28th 1915

Lance Corporal A. G. Curtis, Stoke Bishop. – 1st Batt., Dorsets.
Pte. P. W. Baker, Stoke Bishop. – Signallers' Staff, 3rd Dorsets.

Nov. 11th 1915

Capt. C. B. Greenhill, Dunball, Bridgwater – West Somerset Yeomanry.
Lieut. E. A. Green, Oakhill, Shepton Mallet – North Somerset Yeomanry. (Wounded)
Sergt. N. J. Reynolds, Huntspill, Highbridge – 8th Somerset Light Infantry.
Corpl. R. J. C. Ferguson, Bishop's Lydeard – 8th Somerset Light Infantry.
Pte. E. Westcott, Porlock – 1st R.N.D.H., Mediterranean E.F.

The last three were the local secretaries and visiting experts to their respective districts. The BBJ of 19th March, in the report of Somerset's AGM held at Taunton states that Sergeant Reynolds had been reported missing since the Battle of Ypres last September and it was much feared that he was killed.

June 29th 1916

Kenneth A. J. Moore, Radstock – First-class Boy Telegraphist, H.M.S. *Turbulent* killed in naval battle, May 31st. News of the loss of H.M. Destroyer *Turbulent* with all on board came as a painful shock to Mr H. J. Moore of Radstock whose only son, Kenneth was on board that vessel. Mr Moore is well known in Somerset as a bee expert and the lad will be remembered by many beekeepers as he frequently cycled with his father when on tour. (It was, of course, the Battle of Jutland.)

September 1916

Major E. Elger, D.S.O., "Colinsays," Bruton – 1st Somerset Light Infantry.
Capt. St. John Mildmay, Queen Camel – 60th Rifles.
Coy. Sgt.-Major J. Bindon, 11 Coombe Street, Bruton – 1/4 Somerset.
Trooper V. Davis, North Barrow, Sparkford – Imp. Dorset Yeomanry.

July 19th 1917

Pte. J. W. Owen, Churchill, Somerset – A.S.C. – M.T. for two years a district secretary of Somerset BKA.

February 7th 1918

Gunner E. Jackson, Frith Farm, Wincanton – A Battery, 93rd Brigade, B.E.F. France.

The enormity, sheer horror and the tragedies of war brought upon so many families pale into insignificance the problems with beekeeping. But life had to continue at home despite the exigencies of the Great War.

Sugar became very scarce and many stocks were lost through winter starvation. Under pressure from the British Bee-keepers' Association a limited amount of candy (60 tons) was made available by the Government for autumn and winter feeding. This proved hopelessly insufficient and under further pressure from the BBKA a ration of 8lb per stock was allowed. The coupons, however, were honoured at double their value and thousands of colonies were saved. The allocation, secured by vouchers presentable at local shops continued until 1920.

However, bees continued to die from Isle of Wight disease and all sorts of remedies were tried but to no avail – there *was* no cure in these early days. Colonies in modern hives looked after by experienced beekeepers and those kept in old boxes or skeps on the "let alone" principle all suffered alike; British, Italian, and hybrids appeared equally susceptible to the disease though it was probable that certain strains of bees had greater resistance than others.

During 1915 the northern and north-western parts of the county suffered enormously. In a few years nearly 75% of the members had lost all or part of their bees and in many large areas not a single colony existed. Membership of the Association declined as the disease made its large inroads and many branches, like the bees, died out. That year was an exceptionally good year for honey, and where healthy bees were present record crops were obtained.

The year 1916 was remarkable for the white clover flow, but unfortunately there were few colonies of bees to take advantage of this. Supers were filled in a fortnight at the beginning of August, and those members who had bees secured record crops. Only 290 colonies appeared to be surviving. In districts visited by one expert there remained 221 colonies out of 1,233, in another district 6 left out of 220, and in another 31 left out of 119. Sixteen districts had no bees at all, and the Council advised no one to start bee-keeping during the coming year. Unfortunately the native black or brown bee seemed less able to withstand the disease, and in all probability died out of existence completely or was hopelessly crossed with other introduced races of bees.

The following year, 1917 was also recorded as being, on the whole, excellent. Several large "takes" were reported. Mr E. Walker, the Hon. Secretary for the Glastonbury and Street Branch probably set a record for Somerset, if not for the West of England. One of his stocks of Italian hybrids gave the extraordinary yield of 308lb of extracted honey. The hive was kept permanently on scales, and weighed daily from May 1st to August 4th, during which time it increased in weight by 384lb.

It was most unfortunate that so few members had any bees to take advantage of this excellent season. Mr Bigg-Wither records in his annual report for 1917 that over half the visiting experts were in the Army so that many members did not receive a visit. *"Another of our Branch Secretaries and experts, Mr R. Litman of Castle Cary, has joined the Army. Mr F. A. Buxton kindly undertook the work of Mr F. W. Owen, who is in the Army, and visited members of the Churchill Branch."*

Mr L. Bigg-Wither held nobly to his office as Honorary Secretary during these dire years. The following record of membership shows the continuing loss until 1918, when extinction threatened the Association. 1919 saw "the turn of the tide" and the hope that the disease had spent itself somewhat.

1913 – 500 members	1917 – 118 members
1914 – 345 members	1918 – 115 members
1915 – 235 members	1919 – 257 members
1916 – 163 members	1920 – 275 members

It must not be imagined that the Somerset Bee-keepers' Association did not take all the active measures it could to combat this scourge, for many and varied were the preventatives, cures, and disinfectants used. Even at the Council meeting on January 23rd, 1915, Mr Brewer and Mr Withycombe suggested the following: *"All bee-keepers whose bees have suffered from the Isle of Wight disease are invited to send to the honorary secretary short notes, etc., with regard to their experience of the disease and the effects of any remedies."*

However, on February 24th, 1916, the opinion was that no cure had been discovered and most of the so-called cures were worse than useless. Mr Snelgrove advised *"complete destruction of every diseased colony and thorough disinfection by fire of every diseased colony and thorough disinfection by fire of every appliance."* On March 9th 1916, a letter was read from Mr G. W. Judge, of the Cayford and District BKA, giving details of a proposed Co-operative Re-stocking Scheme, but it was considered inadvisable to attempt any re-stocking in the county during the present year.

On January 18th 1917, a letter was read from the British Bee-keepers' Association giving details of a scheme for re-stocking, but Mr Cowan *"thought it would be better to wait another year until the disease had spent itself."*

Many attempts had been made and various suggestions put at Council meetings to find out the true extent of the ravages of the Isle of Wight disease and what re-stocking was necessary. Colonel Jolly, the Chairman, had even suggested that all policemen and head-teachers should help to make this census. However, it was left to Mr L. E. Snelgrove to do it. He reported that before the outbreak of disease bees had been kept by 1,500 persons in Somerset and the number of colonies was about 8,000. The present number of colonies (1918-1919) was 2,000, so that at least 75% had died out. Previous attempts by the Association had put the loss at 90% in the county. These remaining 2,000 colonies were not immune, as it was reported that 680 of these were diseased. Mr Snelgrove reported *"there were 550 parishes and in 75 of these the bees had died out completely. The 6,000 empty hives were known to be infected and remained a source of potential danger."*

The re-stocking Scheme and Recovery

On February 23rd, 1918, it was decided to start a re-stocking scheme in the summer. It was hoped to arrange for several nucleus-rearing apiaries in different parts of Somerset.

On February 8th 1919 Mr Cowan gave an outline of the scheme. The Government had at last recognised beekeeping as an important rural industry. Dutch colonies would be imported in skeps and also a number of Italian mated queens. Dutch bees were chosen because they bred rapidly and Italian queens were considered to be most resistant to the disease.

Members of the Association desiring bees were required to apply for shares valued at one pound each and they would then be entitled to receive a nucleus colony in rotation. Mr Snelgrove did most of the work in drafting the re-stocking scheme for the county.

Six hives and other appliances were purchased and later a nucleus re-stocking scheme committee was formed. On March 17th 1919, arrangements for the reception of the fifteen Dutch stocks, which were being brought over from Holland during the next month were discussed, and it was agreed to distribute them as follows:

5 to Mr H. J. Grist's apiary at Evercreech.

5 to Mr J. W. Heard's apiary at Yeovil.

3 to Mr J. Rudman's apiary at Bridgwater.

2 to Dr Wallace's apiary at Weston-super-Mare.

The Italian queens were to be re-posted to the applicants by Mr Bigg-Wither.

The fifteen colonies of Dutch bees arrived from Holland on April 13th 1919. Several of these arrived in a very weak state and it was not possible to obtain increase from them until late in the season. The Association also possessed 7 colonies and in November 1919, they owned 48 colonies, 23 at Shepton Mallet, 13 at Yeovil, 7 at Bridgwater, 3 at Weston-super-Mare and 2 at Wells. The sum of £52. 10s. 0d. (£52.50) was paid for the fifteen Dutch skeps of bees and 30 imported Italian queens.

During 1920 Mr Bigg-Wither stated that owing to the very unfavourable season nucleus rearing was only made possible by continuous and heavy feeding. 73 nuclei were sent out at 45 shillings (£2.25) each and queens at 8. 6d. (42½p) each.

Fifty colonies of varying strength were owned and the question of wintering this number was discussed. It was evident that this was the beginning of the end of the re-stocking scheme. Mr Litman (Castle Cary) and Mr C. Harris (Bishops Hull) offered to buy the colonies in their charge and this was agreed. Soon it became evident that the disease was definitely on the wane, and in 1921 bees were being established in many parts of the county, although a clean bill of health as far as Isle of Wight disease could not yet be given. Splendid work had been done in Somerset by the re-stocking scheme, but its labours were becoming less.

The disease had caused what was probably the greatest catastrophe in British beekeeping. Many scientists and beekeepers had given much time in patient investigation and numerous experiments had been conducted. It was known that Dr John Rennie, D.Sc., F.R.E.S., of Aberdeen, *"had been experimenting to find out precisely the nature of the malady and a remedy if possible,"* and in 1919 he and his helpers discovered the eight-legged mite *Tarsonemus woodi* and named by him *Acarapis woodi*. This mite, by the invasion of the trachea of the bee was considered to be the entire cause. Dr Rennie published his findings in his memoirs, "Acarine Disease Explained," but it was left to Mr R. W. Frow of Lincolnshire to devise a cure known as the Irons Method, at one time well known to most beekeepers.

Meanwhile, Mr Snelgrove was pursuing his own line of research. The following two advertisements appeared in the British Bee Journal.:

7th August 1919 - Mr L.E. Snelgrove, Albert Quadrant, Weston-super-Mare, would be grateful to readers who would send him, with brief notes, specimens of live bees (2 or 3 dozen) affected with I.O.W. disease for scientific purposes.

21st August 1919 – I.O.W. Disease – wanted for scientific purposes a badly affected stock. Mr L.E. Snelgrove, Albert Quadrant, Weston-super-Mare.

Chapter 3:
New Rules and Advancement

New Rules and Re-organisation

It was at the Annual Meeting on February 9th, 1920 at The Assembly Hall, Yatton, that Mr Snelgrove moved a resolution *"That this meeting instructs the Council to consider the present constitution of the Association and draw up a set of rules so as to place the Association on a more representative and efficient footing."* He considered the existing organisation was *"very antiquated and inefficient. They in Somerset prided themselves on being in advance of other counties in the business of bee-keeping, but it was about time they reviewed their organisation and made it more effective. They had twenty branches scattered about the county. Instead of electing the delegates haphazardly they should have a sane system. The local delegates should be elected by the branches."*

In December 1920 the new Rules of the Association were passed. The county was divided into five divisions, *viz,* Northern, Southern, Eastern, Western and Central, with a divisional secretary in charge of each, all electing their own representatives. It is important to remember that the Council realised the worst of the disease problem was over and with the growing influx of members, it was their plain intention to build up again in a modern way.

At the same meeting Mr T. W. Cowan gave an account of the conference in Westminster, attended by Mr Snelgrove, Mr Bigg-Wither and himself, to discuss the question of legislation in connection with the Foul Brood Diseases Bill, at which the Board of Agriculture had invited all the Associations in the country to send delegates to ascertain the need or otherwise for legislative action – 99 of those present voted for, and 1 voted against.

The year 1920 was long remembered as one of the worst years for beekeeping. The season opened well, but at the end of May the weather became so unfavourable that practically no further honey was stored; "hunger swarms" were very frequent during August and wintering was a problem.

By 1921 Mr L. E. Snelgrove had done considerable research work in connection with the so-called "Isle of Wight" disease, and for a thesis on this subject had been given the degree of M.Sc. On August 11th 1921, there appeared in the British Bee Journal the following advertisement: *"Somerset Re-stocking Committee having decided to close down its apiaries this year, has a number of Nuclei and Stocks for sale, specially selected strain, also hives and appliances. What offers? L. Bigg-Wither, Birdwood, Wells".*

It was on February 23rd 1922 that Mr L. Bigg-Wither resigned as Honorary Secretary because of his removal from Wells, and Commander R. D. Graham was elected in his place. In August Mr Bigg-Wither was presented with a clock at the summer meeting of the Council at Cannington Farm Institute.

Prominent Members die, but others rise to Prominence

At the Council Meeting held at the Railway Hotel, Bridgwater, on April 3rd, 1922, it was reported that Lieut.-Colonel Jolly, Chairman of the Association for almost ten years, had passed away. Mr Snelgrove proposed a resolution of condolence.

It was then proposed by Mr Withycombe and seconded by Mr Bradbury that Mr L. E. Snelgrove be appointed Chairman of the Association to fill the vacancy. The Honorary Secretary read communications from Mr T. W. Cowan and Mr L. Bigg-Wither supporting this action. Mr Snelgrove, in accepting the position, referred to the honour done to him by the Council and stated he would do all in his power to justify this confidence placed in him.

Another loss to the Association at this time was that of Samuel Jordan of Bishopston. He had become a First Class Expert in 1897 and was Chairman of the former Association for six years until it was wound up in 1906. Mr Jordan had, in the last century, taught at Street, Milborne Port and Princes Risborough until becoming headmaster of St Philip's Boys' School, Bristol. He had been a tower of strength in the advancement of this Association and its predecessor until ill health had forced him to withdraw. He was elected to honorary life membership in December 1920 in respect of 45 years of loyal service to the two Associations.

It was in 1923 that Richard Beck (1858-1945) began to examine bees for Somerset beekeepers. His first notes were made about Acarine (I.O.W.) disease in 1921, although he had been

Richard Beck 1858–1945

interested in this disease for many years. For over twenty years he faithfully recorded his notes and work with those beekeepers to whom he taught dissection and the use of the microscope for bee disease diagnosis. The Somerset BKA owes a great debt of gratitude to this splendid microscopist for his patient and enduring work during this period until his death in 1945.

Another member who came into prominence at this time was Dr C. R. Killick of Williton. He too, was a competent microscopist and was greatly interested in the acarine mite. He was a member of the Apis Club and his letters were published in Bee World. This was the journal of the Apis Club, subscribed to by the elite, the educated and the upper echelons of beekeeping, both in the United Kingdom and abroad. Hence, Dr Killick became widely known.

Recovering from the vicissitudes of the Great War and the ravages of I.O.W. disease, the Association now experienced a period of expansion, both in its activities and in its membership. The team of visiting experts played a key part in this growth. Those prominent over the next few years were Messrs Hawkins, Litman, Pritchett, Trump and Withycombe.

Eastern Division, Mr Hawkins and Downside Abbey

For more than thirty years Mr E. G. Hawkins was variously expert for the Eastern Division or certain of its branches and also Hon. Secretary at branch or divisional level, rendering invaluable service without holding high office at county level. He was a skilled engineer engaged on work concerning railways, necessitating travel. At weekends and holiday times he was hard at his expert's work. If operating at some distance from home, he would catch a train from nearby Frome station, taking his bicycle with him, to Radstock for instance, and complete his pre-arranged itinerary in that area on two wheels.

He found it an interesting and rewarding diversion from his professional work, taking

Left: Mr E. G. Hawkins employing a simple means of elevation, and, right: at work in a member's apiary.

Photos courtesy of Miss S. Hawkins

him to interesting places and meeting people from other walks of life. Bees were kept at Downside Abbey, Stratton on the Fosse. Many young monks tried beekeeping over the years, and in Father Aidan Trafford's time consideration was given to making honey production a major activity. Fr Trafford was aware of the success being made by the young Brother Adam at Buckfast, with its apiaries over Dartmoor, and he wondered if the Mendips could be exploited in the same way. Mr Hawkins advised against it, the conditions being different, but beekeeping continued in a small way and for a time Fr Trafford was a member of the struggling Radstock branch. Part of the operation was the management of some hives in the unique wooden thatched bee house, illustrated. This was situated outside the monastery end of the Petre swimming pool. By 1943 the thatch had been replaced by tiles and the building converted into a summer house. It came down in the great storm of 1987.

Photo courtesy of Downside Abbey Archives

Each summer Mr Hawkins played host to large gatherings of beekeepers at his home, Oaklands, Rodden where there was a large garden, orchard and apiary. Many famous beekeepers including William Herrod-Hempsall were guests at such gatherings. Mr Hawkins liked to purchase a ticket for the Irish sweepstake. On the first of such occasions, which coincided with the race, there were so many motor cars outside his property, it was rumoured in the locality that he had won and that celebrations were proceeding!

Miss Sylvia Hawkins demonstrates.
Photo courtesy of Miss S. Hawkins

Experimental Apiaries

In 1923/24 the formation of experimental apiaries was considered. These were to be set up in such places as Cannington and Long Ashton. It was then proposed to purchase French bees, and some hives and appliances. Those forming the Experimental Sub-Committee were Mr L. E. Snelgrove, Dr C. R. Killick, Mr J. Spiller, Mr L. Bigg-Wither, Mr W. Withycombe and Mr W. West (Honorary Secretary).

The expressed purpose of these apiaries was to test new theories and methods, and to demonstrate their value to the members. Two questions were specially exercising the minds of beekeepers:

(1) The suitability of French bees, which were said to be free from Acarine disease.

(2) Whether in this country there would be any advantage in adopting a larger frame than the British Standard.

Mr Withycombe took charge of the Cannington Apiary and Mr Bigg-Wither the one at Long Ashton. Mr F. W. Moore later took charge at a more central apiary at Yatton.

The Annual Reports of 1924 and 1926 contain much interesting information about these experimental apiaries and the results they revealed. One experiment was the treatment of combs affected with foul brood. The apparatus was installed at Fosseway, Durleigh Road, Bridgwater, and the treatment, with formaldehyde, carried out under the supervision of Mr Rudman.

Commander R. D. Graham resigned as Hon. Secretary on February 28th, 1924 and Mr William West was appointed. For sixteen years "William" held this office, during which there were many troublesome and conflicting episodes in the history of the Association.

"Mr West did a great job ... and did it well." The report of the season 1924 stated that the year had been one of the worst ... if not the very worst on record owing to continued wet weather. The Frome district, strange to say, reported a surplus. and entries were reasonable at the Taunton Show.

Honey entries at Taunton Show, held in conjunction with the Bath and West Show, 1924. *Photo: B.B.J.*

During 1925 branches of the Association were re-formed at Radstock (with Midsomer Norton) and Ilminster. It was reported that Mr C. Tite of Taunton had presented the Association with some valuable and very interesting items, including photographs, slides, Italian plates, glass cases, etc.

Death of T. W. Cowan: the end of an Era

The year 1926 marks the date of the death of Thomas William Cowan whose name, familiar amongst progressive beekeepers throughout the world, will ever be remembered in beekeeping. His death was an irreparable loss to the Somerset BKA. Not only was he its President, he was also President of the British Bee-keepers' Association. It was at a Council meeting held at Bridgwater on June 19th, 1926, that the repercussions were considered. In view of the great work done by Mr L. E. Snelgrove it was unanimously agreed he would make an ideal President and deserved the honour.

At a subsequent meeting at Cannington on July 14th, 1926, Dr Killick suggested that Mr Snelgrove be asked to accept the Presidency and Dr Wallace the chairmanship, and that this should be submitted to the General Meeting at Long Ashton on August 21st, 1926. It was there on the proposition of Dr Killick, seconded by Mr J. C. Morland, that the recommendation was made that Mr Snelgrove be appointed President of the Somerset BKA and this was duly confirmed at the Annual Meeting held at the Mermaid Hotel, Yeovil on February 24th 1927.

Dr John Wallace, O.B.E. became Chairman and he continued in this office until he resigned through ill health in January 1947, in his 91st year. He had also been President of the Northern Division since 1924 and a Vice-President of the British Beekeepers' Association.

Some Politics and Theatre: a Member is expelled from BBKA

It would seem that within a few years of the inception of the BBKA differences of opinions caused divisions, and the frequently recurring question was whether the County Associations received sufficient consideration and whether the central body should not do more for the common good.

By the 1920s it appeared that the BBKA was seriously in decline and not effectively fulfilling its role as the national head of a large group of affiliated local associations. Both Scotland and Ireland had broken away and formed their own groups. In England several county associations had seceded, most seriously the counties of Kent, Surrey and Sussex who had between them, formed a rival association. A faction in Somerset was all for following suit. Dr Killick felt strongly that the Somerset Association should remain loyal, building and improving on what Cowan and his associates had established on such sound foundations in their time.

Towards the end of 1925 Dr Killick managed to secure, not only for himself, but for Mr W. H. Ashwin, the BBKA delegate, an invitation to visit the elderly Mr Cowan at his home in Clevedon. The prime purpose of their visit was to see Mr Cowan's microscope built by him many years ago from castings provided by the firm of R. & J. Beck. Cowan, fearing that the conversation might turn to the parent body of which he was president, asked that Messrs Snelgrove and Beck be included in the party. Thus, what was intended as a private visit turned into a delegation. Mr Snelgrove, Killick admits, had cautioned him against saying anything about the BBKA that could hurt Mr Cowan's feeling about the Association in which for so many years he had taken such a practical interest. Dr Killick denied having any intention in this direction.

There appears to have been a showdown at the annual general meeting held at Bristol on 27th February 1926. Dr Killick brought forward a resolution: *"That the BBKA should be reconstructed on lines which will bring affiliated and all other associations interested in beekeeping in the country into closer contact with the BBKA, and that this involves a basic reconstitution of the Association."* His friend Mr W. H. Ashwin seconded the resolution. Mr Eldred Walker (Bristol Branch) greatly amused the meeting with his outspoken and unsparing condemnation of the BBKA and all its ways. The resolution was passed unanimously. Bee World reports a scathing attack on BBKA for *"doing nothing, it being ill adapted to modern thinking"*. W. Herrod-Hempsall, Secretary to the BBKA, was present at the meeting as guest speaker and referred to a report by the BBKA. He gave no idea of the nature of the report, but said that the objectors would find *"some of their bones stolen"* when it was circulated.

Then the Western Division passed a resolution to the effect that the 1927 Annual Meeting should consider the desirability or otherwise of withdrawing from affiliation to the BBKA. Mr J. Spiller, (Taunton Branch) said that the Association was of no value to them except in respect of examination fees. It was antiquated and they wanted something different. Dr Killick reiterated his view about the need for reform, urging that all should become members of BBKA so as to secure national representation.

The proposition to leave the BBKA was not seconded, and it was then agreed to remain within and draw the BBKA's attention to the need for reforms. Shortly afterwards Dr Killick attended the 1927 BBKA Annual Meeting in London instead of Mr Ashwin. At this meeting he proposed three motions on behalf of Somerset BKA. One failed to find a seconder and the other two were defeated.

The annual report for 1927 contains this reference: *"The matter of the reconstitution of the BBKA has been before all our meetings for some years. It has taken up considerable time at our Annual Meetings as well as at our Council Meetings. My information is that those members who were sufficiently interested to follow the matter are now tired of this hardy annual and the time has come when this may well be left out of our programme and our energies concentrated on the great amount of work to be done in the county itself."*

It is unfortunate that owing to lack of documentation one cannot be precise about the whole of this intriguing series of events, but the intrepid Dr Killick appears not to have tired of the matter and continued to pursue his quest for reform with dogged persistence.

Whatever ensued was apparently regarded with great disdain and irritation by the Council of the BBKA. The following year (1928) the saga culminated in Dr Killick's expulsion as a member of the national body of which he was a Vice-President.

In consequence, Mr Ashwin resigned from the Somerset BKA. Dr Killick resigned as Vice-Chairman of Somerset BKA and as Hon. Secretary of the Williton Branch. By 1928 membership of the Williton Branch had fallen from 34 to 20 and by 1929 had disappeared from the year book, it having amalgamated with the Taunton Branch. No explanation is given and one is obliged to draw one's own conclusion. Dr Killick joined the Taunton Branch and remained a member of the SBKA Council.

Dr Charles R. Killick of Williton, with grandson John, circa 1934. *Photo courtesy of Mrs. J. Killick*

This unfortunate matter was eventually resolved, probably through the good offices of Mr Snelgrove. BBKA Council minute, item 11. of 28th January, 1950 reads as follows: *"It was proposed by Mr Wadey and seconded by Mr Thornton that the Council shall hereby solemnly dissociate itself from the action of the Council of 1928/29 in expelling the two Vice-Presidents, Dr C. R. Killick and Mr Herbert Mace, and shall hereby instruct the Secretary to expunge from the Minutes the record of that shameful action. This was passed with acclamation after supporting speeches from the President and others."*

A contemporary report states that the BBKA Council heard with sympathy requests from various places to expunge the record. It further states that the action was taken because these two members insisted on the rules of the Association being observed! Thus although it was eventually admitted that Killick's harsh and unjustified treatment was a misjudgement by the BBKA Council, exoneration came too late for Dr Killick as he died in 1941. Herbert Mace, an Essex beekeeper, prolific activist, reformist and author, who was expelled the following year, 1929, was still alive.

1928 saw the inauguration of an Insurance Scheme to the liability of £1,000 for beekeepers and Mr A. D. Turner, the Horticultural Superintendent, at Cannington Farm Institute introduced a scheme for better marketing and advertising of home produced honey. It was stated that the Merchandise Marks Act had become an accomplished fact and

New rules and advancement 43

the country of origin had to be stated on imported honey. Mr Turner gave the Association valuable assistance in every way for a great number of years. Membership now stood at 602.

The year 1929 began with an extremely cold spell, followed by one of the most remarkable summers on record … never it seemed had there been such a prolonged drought. With the absence of rain, conditions became too dry and the nectar flow fell away to nothing. Mr Snelgrove wrote, *"it was one of the worst, if not the worst season in living memory."* In August swarms were dying of starvation, and many good colonies perished by starvation during the following winter. Losses were also severe the following spring.

1930 was also a poor season … *"much hope but little honey"*. Membership was now 633. Both acarine and foul brood diseases showed an increase. The Eastern Division was reduced to two branches consequent on Bath being placed with Bristol in a new North-Eastern Division. Mr E. G. Hawkins was doing much hard and patient work in sustaining the Eastern Division.

Apart from being the visiting expert for the Southern Division, Mr A. E. Trump of Alhampton, near Castle Cary, was Hon. Secretary to two of its branches. In addition to this, he was now the delegate to the BBKA. The recent acrimony fading into the past the Association continued to make useful input for the benefit of the craft. At the monthly meeting of the BBKA Council in London on 19th November 1930 he suggested that BBKA take up the matter of grading honey on the lines already inaugurated in Somerset. He gave details of what had been accomplished by this association but after a thorough discussion it was resolved that *"as the markets branch of the Ministry of Agriculture and Fisheries already had the matter in hand, under the National Mark scheme, it was considered wise not to start what might be considered a rival scheme…"* It was

W. Herrod-Hempsall judging honey at Bristol.

resolved that the best thanks of the Council be given to the Somerset Association for the work they had done in regard to a grading scheme. This was carried unanimously. The following year, 1931, SBKA wrote a letter suggesting an examination certificate for honey judging. This was taken up, and illustrates Somerset's continuing involvement in national affairs.

The SECOND BRISTOL HONEY SHOW

LAST YEAR'S Attendance: 10,000 in TWO DAYS.

Wednesday-Thursday, November 13-14th, 1929

THE Somerset Beekeepers' (Bristol Branch) Association are holding their **SECOND SHOW** in conjunction with

Bristol Horticultural & Chrysanthemum Show

The MOST ATTRACTIVE SHOW IN THE WEST OF ENGLAND.

=== A NEW IDEA! ===

THE BRISTOL SILVER QUEEN, a beautiful and attractive Model of a Queen Bee, in Silver, on Ebony base, together with a **£5 MONEY PRIZE** is offered in Class 1 for the best Four Sections and Four 1-lb. Bottles. This Prize is within reach of every Beekeeper in the United Kingdom. Small or large, Entrance 2/- each entry.

GOOD PRIZES including CUP, ROSE BOWL, MEDALS, &c.
OPEN & LOCAL CLASSES.   GLOS. & SOM. COUNTY CLASSES.

Particulars from Show Secretary—E. LEWIS, "Sonoma," Abbots Leigh, Bristol.

L. E. SNELGROVE,

WHO IS ADOPTING A UNIFORM TYPE OF HIVE IN HIS APIARY, WISHES TO DISPOSE OF A NUMBER OF GOOD

SECOND-HAND

HIVES AND STANDARD SIZE SUPERS

HIVES	from 10/-
SHALLOW FRAME RACKS	,, 1/6
SECTION RACKS	,, 1/6

May be viewed by appointment. :: Purchasers to remove.

ADDRESS :—

14, ALBERT QUADRANT, WESTON-SUPER-MARE

Right: Mr Snelgrove's advertisement from the SBKA Year Book of 1930

CHAPTER 4:
The 1930s, but another World War follows

Good and bad Seasons

The last two seasons were bad enough, but the year 1931 proved to be one of the worst years on record. Not only was there a dearth of honey, but artificial feeding had to be resorted to throughout the summer and many of those who failed to do this sustained losses. Diseases were prevalent, particularly foul brood and acarine. Frow's treatment for the latter proved effective, but only three members sent combs affected with foul brood to the station for disinfection.

The SBKA Council recommended that we join the South Western Counties Federation comprising Somerset, Devon and Cornwall. The function of Federation Council generally was to discuss matters of insurance, honey grading, shows, conferences, co-operation and adoption of common schemes, combined summer meetings and other matters of common interest. The Federation already arranged a course each year at Seale Hayne College in Devon, but a similar one arranged at Cannington the previous year had to be abandoned for lack of support.

Then came a fourth poor year. In 1932 not a single good "take" was reported. Many members lost all their stocks of bees through starvation and disease. Foul brood and nosema caused serious losses, but Frow's treatment against acarine continued to prove effective. Mr Snelgrove commented on the disagreeable surprise of some members at seeing a plentiful and continuous supply of bottled honey labelled "Pure English" in certain shop windows. Strong doubts were expressed as to its origin and the SBKA Council, suspecting fraud, took the matter in hand.

These four successive poor years, with little honey and much evidence of disease caused some members to give up in despair and membership declined somewhat. The beekeeping scene was set against a backdrop of a gloomy situation worldwide. This was the time of the "Great Depression" – the economic slump that followed the 1929 Wall Street crash. Companies went into liquidation, millions of people were thrown out of work and economic conditions were harsh. In 1932, as part of its economy campaign, Somerset County Council was obliged to reduce the Association's annual grant from £50 to a mere £10. This necessitated discontinuance of practically all the free experts' visits in the autumn of that year and a curtailment of the lecture programme that had also received funding.

At last came the long awaited upturn. For Somerset beekeepers 1933 was one of the most favourable for years and much surplus honey was taken throughout the county. The largest quantity taken from one stock was 230 lb. This was taken from a hive situated in the garden at Crackmore, Milborne Port, near Wincanton, belonging to Mr A. G. Pilkington and worked by his friend Mr Snelgrove using his non-swarming method. The wholesale price for honey recommended by council was 1/6d (7½ p) per lb. Experiments continued with Mr Snelgrove's non-swarming methods and with a satisfactory conclusion it was hoped that he would publish a book on the subject. The experimental apiary where the method was tested was moved from Yatton to Wembdon, Bridgwater, to a site that proved unsatisfactory.

1934 was a splendid year! There were successful honey shows accompanied by the perennial complaint still applicable today, that not enough members supported them.

The £50 grant from Somerset County Council was reinstated. Some of this grant was spent in training members in the use of microscopes, primarily to assist them in the detection of acarine mites and nosema spores. The training was done by Mr Richard Beck.

Mr H. J. Rudman's Apiary at Bridgwater. *Photo: W. Herrod-Hempsall*

Mr H. J. Rudman, a longstanding Bridgwater Branch member, had done good work for the Association over a great many years. *"Mr Rudman"*, says Herrod-Hempsall, *"loves his bees, and also his garden. Therefore, in order to keep the latter as vast as possible, the flat roof of his workshop, honey house, etc. provides the room for his extensive apiary."* It was in one of these buildings, to the rear of Fosseway, Durleigh Road, Bridgwater, that he housed the Association's disinfecting station. Foul brood was still prevalent, but as the apparatus was little used it was put into store at Taunton in the care of Mr Clarence Harris.

In some districts 1935 was even better than 1934, lime trees and clover yielding well. The stocks in the experimental apiary were sold, thus bringing to an end over ten years of useful work. With successful shows, a more extensive programme of lectures and demonstrations and an increasing membership, William West the Hon. Secretary considered the Association to be in a healthy and prosperous condition.

The 1930s, but another World War follows

1936 was one of the worst years on record but membership continued to grow. 1073 expert visits were made and of these 472 were paid for from funds, 336 from the County Council grant but for 259 visits no payment was made. Visiting experts were still playing a key part in the Association's work. This year the Royal Show was held at Ashton Park, Bristol. Several members were involved with the Hives and Honey section, including Messrs Lewis, (Abbots Leigh) and Withycombe who were the judges. There were classes restricted to members of Somerset BKA, but unfortunately, due to the poor season, not everyone who entered was able to put the anticipated entries on the bench.

Although 1937 was an improvement on the previous year, it was nevertheless disappointing owing to spring losses and a poor summer. Not only was honey scarce, funds had diminished considerably and the year finished with a balance of only £2. 8s. 7d. (£2.43) Some members showed appreciation by offering to double their subscriptions and a "whip round" resulted in over £10 being collected. Things continued through the thirties in much the same way, with good seasons and bad. William West's long period as Hon. Secretary came to an end in 1938 and Mr Pardoe was elected to succeed him. The death occurred of that longstanding, dedicated worker and one-time Secretary, Mr Lovelace Bigg-Wither. In his memory the Association established a library consisting of some of his books, supplemented by gifts from Messrs Snelgrove and Beck. Mr Hermon of Weston-super-Mare was Hon. Librarian, a post he held for many years.

Members at their summer meeting at Long Ashton Research Station, Bristol in 1939; Mr L. E. Snelgrove prominent at the front; Miss M. D. Bindley to his right. *Photo courtesy of Ivor Davis*

Wartime Beekeeping

Beekeeping, for most members, was a welcome diversion or escape from the burdens and anxieties of life, so no wonder there is no mention of the gathering storm clouds: the expanding Third Reich, the restoration of German military power, the persecution of Jews or the re-armament by Great Britain and France and the threat of war. So although war was declared in September 1939 it receives no mention in the Hon. Secretary's report for that year.

By 1941 the profound impact of World War II was very evident, for many difficulties were being experienced in running the Association. Mr Pardoe, secretary for only two years, resigned prior to being called up for the Navy and was replaced by Mr Batty. Mr Whittaker, Hon. Secretary of the Central Division was called up, as was visiting expert, Mr Fred Sparks of Somerton Branch. When key figures left a group its very existence could be placed in jeopardy. In Somerton, Mr Courtney Locke took over, but he was obliged to move to Surrey where he was engaged in war work. He later returned to his old home in Somerton and resumed his beekeeping. Many members were called up to join the Forces, ARP or Home Guard and older members undertook extra work also. Travel facilities were uncertain and petrol rationing put a severe strain upon the organisation of the Association. However, the value of beekeeping received Government recognition. The Ministry of Food considered honey production and blossom pollination an essential feature of the home-grown food campaign and made allocations of sugar for feeding bees. Despite the restrictions imposed upon the working of the Association by wartime conditions, activities such as lectures and demonstrations were greatly increased and by 1942 membership exceeded a thousand. Beeswax was in very short supply and members were urged to save every little bit possible. Captain Stoker became Acting Hon. Secretary when Mr Batty went off to serve with the Royal Engineers, rising to the rank of Lieut. Colonel.

The Foul Brood Diseases of Bees Order, 1942, became effective on March 25th, 1942. Under this Order, each Agricultural Executive Committee became responsible for administering the provisions of this long awaited Act. Somerset BKA worked in close collaboration with it and 28 experts volunteered to examine hives for foul brood. 41 hives were confirmed as infected and at least six other cases were dealt with unofficially. In one case the owner refused to burn his colonies and disinfect the hives as directed. He was therefore summonsed under the Act. After receipt of the summons he decided to carry out the directions but was fined a nominal sum of £2 for failing to carry out the Order within the specified time.

In the succeeding year the experts, working under the auspices of the War Agricultural Executive Committee, visited 603 apiaries, examined 2,547 stocks, travelled over 4,500 miles and expended some 1,336 hours on the operation of the Foul Brood Order. Brood disease was found in almost every part of the county, especially in the areas of Exmoor, Bridgwater and Bath resulting in the destruction of 119 colonies. The supply of petrol to enable inspections under the Foul Brood Order was controlled by the County War Agricultural Executive Committee. There was no objection to an Appointed Officer using

THE BEAR HONEY COMPANY LIMITED

Invite members of the Somerset Bee-keepers Association to send them their offers of

HONEY

Write to the Managers:—

L. Garvin & Co., LTD.

SEYMOUR HOUSE.

TOTNES ∴ DEVON

Tel. 2261.

SUPPLIES FOR 1944.

Owing to war conditions the quantity of bee supplies able to be made is far below normal, but in order to increase output we are concentrating production on to fewer lines, and are doing all we can to distribute what is available equitably amongst our customers.

You can help by ordering your supplies in good time, and by keeping orders down to bare essentials.

Current prices gladly quoted on request.

☞ Send 4d. for 1944 Price List ☜

BURTT & SON, Stroud Road, Gloucester

MAKERS OF QUALITY BEE SUPPLIES.

Wartime advertisements from the SBKA Year Book of 1943

this petrol when working as a visiting expert if during the same journey he inspected hives for foul brood. Petrol was also allocated to experts to perform their duties, to enable members to attend demonstrations and to move bees for pollination purposes.

The 1944 season was notable for the complete failure of the main sources of nectar: clover, lime and heather. The average yield throughout the county was probably not much more than 10lb per colony, but as always, there were exceptions. In the Bridgwater area some beekeepers recorded surpluses of over 50lb, this being attributed to early fruit blossom and late flowering clover.

This was the year in which the reconstituted BBKA commenced to function. The SBKA Council had given very careful consideration to the draft constitution and bye laws of the new body. Council felt that too much power was vested in the Central body and proposed a long list of amendments. All, with one minor exception, were accepted and incorporated into the new byelaws.

In order that the BBKA should have sufficient funds to function adequately, a capitation fee of 6d (2½p) was levied on all affiliated associations. In order to accumulate a reserve for SBKA the subscription was increased by one shilling to six shillings (30p), except for members who did not require an expert's visit and these paid 3/6d (17½p).

In 1944 Somerset County Council made an appointment that would benefit beekeepers in Somerset for the next 25 years. It appointed Mr A. C. Rolt as County Beekeeping Instructor and he took up his post, based at the Somerset Farm Institute, Cannington, on 1st August. Arthur Rolt was a native of Harpenden, Hertfordshire and received his training in apiculture under the renowned Mr W. Herrod-Hempsall at Luton. For 21 years he had been employed as apiarist at Rothamsted Experimental Station before moving to Somerset.

There is no mention of a "bumper season" for any year during the war, and the year 1945, when hostilities came to an end, proved to be a bad one. Mrs Marie Colthurst, who lived at Wembdon, Bridgwater, began producing her Crop and Weather Reports, an annual feature of the SBKA year book for very many years. She reported a dry, warm and sunny March and April with early blossom abundant. Then on 28th April the favourable weather broke. Rain and sleet gave way to snow and was followed by a severe frost. All the blossom and much of the fruit throughout Somerset were lost. Many colonies that grew to strength early starved. Sugar was in short supply and allocations for beekeepers were too little and too late. Other colonies swarmed profusely in June and the weather was so poor that beekeepers were unable to inspect their stocks.

CHAPTER 5:
The early Post-war Years

A Period of Expansion and Change

Although the war had ended it was followed by a long period of austerity. The winter of 1945-46 was severe, as was the summer that followed. Many stocks perished but the losses would have been even higher had it not been for the prompt action of BBKA in securing grants for extra sugar, not only in June 1945, but also an increased allocation for autumn feeding. Membership increased from 1488 to 1582. A summer outing to Brother Adam's apiary at Buckfast Abbey was well attended.

At Bath the phenomenal success of the re-organised Bath Honey Show, including the Eight Counties Championship attracted entries from all over the Kingdom. Mr E. J. Betts, the Hon. Secretary of the Bath Division and his committee were highly praised for their efforts. This event, in conjunction with the West and South West Counties Beekeepers' Rally, held in the Pump Room, continued for several years.

Bath Honey Show, 1948. The Trophies with Show Organiser E. J. F. Betts (left) and Lecture Organiser A. C. Rolt (right). *Photo courtesy of H. Teal*

The South-Western Federation was dissolved by mutual agreement of the member associations and attempts to form a federation with the counties of Wiltshire and Dorset were unsuccessful. Even in these golden post-war years there were exceptions to the rapid move forward and they illustrate the point that throughout the Association's history, local groups rely greatly on the dedication and expertise of officers and committees. The term *"dead areas"* appears in the minutes of the Council meeting held at Bristol on 9th February 1946. Blagdon and Wedmore were referred to as backward areas and Captain Stoker suggested that they might be helped by drawing on lecturers from the pool.

The winter of 1946-47 was one of the worst of the century and was accompanied by frequent power cuts and bread rationing. There was a heavy toll of colony losses again, many being attributed to shortage of winter stores. The summer of 1947 provided a great boost to the craft, having been the best for many years, the favourable weather continuing until the end of August. It was observed that bees at Cannington, at least five miles from the nearest moor, gathered considerable quantities of ling honey.

A benevolent fund was established for the benefit of members who had suffered misfortune arising directly from their activities in the management of their own bees or when in voluntary service on behalf of the Association. Membership rose to 1587 in 1947 compared with 1582 in 1946, and it was observed that 2500 beekeepers in Somerset applied for a sugar allocation for feeding bees. This meant that for every three who were members there were two who, for some reason or other, were not.

Captain Stoker was still acting as Hon. Secretary, but jointly with Mr Batty on his return to civilian life. However, ill health caused Batty to give up and Leslie Hender gave Captain Stoker some assistance in running the Association. Also, in 1945 Miss Bindley was appointed as Education Secretary, relieving Captain Stoker of a large portion of his workload. Later Mrs Lilian Hender became Minuting Secretary. The management and day to day running of such a large association with so many Divisions and Branches was an onerous and time consuming task and this is when we see the administrative work being spread more widely for the first time.

Dr John Wallace, O.B.E., who in January had relinquished the office of Chairman after twenty years of service, died the following month in his 91st year. He was President of the Northern Division and a Vice-President of BBKA. Dr Wallace was succeeded as Chairman by a much younger man, the 67 years old Major John Lintorn Shore, DSO, OBE, JP. He lived at Whatley House near Frome and was President of the Eastern Division. Sadly, and regrettably for the Association, Major Shore, a regular officer in the Cheshire Regiment who for nearly thirty years was actively associated with beekeeping in Somerset, died suddenly in December 1947 after less than a year in office. He was succeeded by Mr Leslie Hender, and Mr Betts, who had made such a mark in Bath, became the new Hon. Secretary.

1947 was the year when the price control on home produced honey was lifted and the BBKA recommended a price of 3/6d. (37½p) per lb, *"properly put up and labelled"*. This was

The early Post-war Years 53

also a notable year for visiting expert Mr George Jenner, a keen and well qualified member from near Chard who was appointed as the County Bee Instructor for Devon. In Somerset our own Instructor, Mr Rolt was reported as having paid 53 advisory visits, written 591 advisory letters and made 166 advisory telephone calls. At the Somerset Farm Institute 55 students were voluntarily taking the beekeeping course.

By 1948 there were no fewer than 1659 members. Thirty candidates attempted the BBKA Preliminary examination, (presently called the Basic), and successful ones included such well known names in later years as F. Buckley, L. Gulliford and R. W. Sawyer. Twelve candidates attempted the Intermediate examination, six attempted the Senior and two were examined to become BBKA lecturers.

During the war years, not only did membership surge, but also the proportion of women increased considerably. As long ago as 1890, writing in the British Bee Journal on whether beekeeping is a suitable or desirable occupation for women, Henrietta Buller states *"there is no reason why any woman of moderate strength and intelligence should not be able to take charge of an apiary of from 30 to 50 colonies with very little assistance"*.

Management of the Association, however, was male dominated, as were most walks of life in these times. True, the first president was a woman, and Miss Bindley had played an active role since her arrival in Somerset in the 1930s, but in the late 1940s many more women played a prominent part. In his report for 1949 the Hon. Secretary writes: *"Our thanks are due to our ladies – Mrs Colthurst, whose excellent weather articles are so carefully compiled. Miss Bindley, for her enthusiastic work as Education Secretary. Mrs Turner (wife of*

A. C. Rolt demonstrating at a garden meeting in Minehead.   *Photo courtesy of J. C. Newcombe*

the Horticultural Superintendent, Cannington Farm Institute), Mrs Hender of Taunton, Mrs Gracey of Bristol, Mrs Moore of Somerton and the host of ladies who grace the various Branch meetings, supplying refreshments and generally helping on the work of the Association".

In 1949, beekeepers experienced their best season for many years, it being remembered for an abundant surplus, well-tempered bees, less swarming and ideal weather for demonstrations. Local honey was offered for sale in all parts of the county at prices ranging from 2/9d (14p) to 6/6d (32½p) per pound, but many samples displayed in shops should never have been offered to the public. The Hon. Secretary reported that more care should be taken in preparation, in some instances the honey appearing as though it had been pressed out, and bottled without straining. *"Members who continue on these lines"*, he went on to say ***"will find housewives will refuse their honey in favour of the imported honey, which usually is better packed and labelled, and also is cheaper and looks more appetising."*** Membership peaked at 1744. It is frequently said that many people were merely putative beekeepers desiring eligibility for the extra sugar allowance. There is some truth in this, but the position is probably overstated as membership started to decline three years before the allowance was discontinued. One cannot mention 1949 without referring to the death of that pillar of this Association and of its forerunner, Mr W. A. Withycombe. Had the three-year rule existed in those days here was a man who would have made a fine President and who well deserved the honour.

From 1948 to 1952 the seasons became increasingly good, followed by 1953 and 1954 which were poor. To give details of every year would make this a weighty tome and become rather boring, but the aim is to show the variation from year to year and major developments and trends as they occured. In 1952 John Spiller, the prominent Taunton member, published his little book on "Bee Houses".

John Spiller

Over the Peak

Between 1950 and 1951 the Association suffered a decline in membership of over 200. The winter was appalling with severe losses amounting to hundreds of colonies and in 1951 there was little surplus honey. It was felt at the Annual General Meeting that the fall in membership was due partly to these severe losses and many members felt that they could not dispose of their surplus honey at a price that would repay them for their labour and the ever-rising cost of equipment. The fall in membership resulted in less income and a declining bank balance. By the end of 1953 this had fallen from £171 to £100.

The early Post-war Years

Once sugar ceased to be rationed in 1953 the Association operated a cheaper sugar scheme. It was administered by Mr Sawyer, the Vice-Chairman. Surprisingly, not all branches took advantage. The sugar, which was from cane, was imported from Queensland, Australia and purchased in bulk direct from the docks. From there it was delivered by road in 160lb sacks to any convenient address and a member not requiring this amount could arrange to share with another. The sugar was straw coloured, pleasantly flavoured and taken readily by the bees which by all accounts wintered well on it. Many members saved the cost of two years' subscriptions through this scheme and the sugar was also useful in the kitchen.

The 1954 season was described by both Messrs Snelgrove and Rolt as *"disastrous"*. An article by Mr Snelgrove in the subsequent year book extolled the benefits of using candy as a winter supplement and described how to make it. Despite the poor summer the Hon. Secretary reported that 89 demonstrations and talks to be given by the Association's experts had been arranged, though some were abandoned due to the weather. Mr Rolt gave 52 lectures and 29 demonstrations. He also gave 38 talks and demonstrations in schools.

Membership continued to fall in 1955 as did the Association's reserves. Members were implored to find Year Book advertisers to ensure this essential annual publication could continue to a standard worthy of the Association. The Langport Branch of the Somerton and District Division and the Martock Branch of the Southern Division both ceased to function.

The Apiary and Hut at Heatherton Park

It was on 17th November 1954, that the Chairman of Taunton Division announced to a divisional committee meeting that Mr Harris had offered a parcel of land at Heatherton Park for use as an apiary site and a shed for divisional use. He further stated that he would thank Mr Harris and obtain further details. Mr Clarence Harris, affectionately known as 'Gassy' because he was for many years, the manager of Taunton Gas Works, was a long standing member of the Taunton Division and had played a leading part in the work of SBKA. Heatherton Park, between Taunton and Wellington, owned by Mr Harris, was one of the many large properties in the area which he bought and converted to flats. During the war it had been a school and it is believed the hut, given with the apiary site, formed two classrooms. Signs of a partition remain visible today.

The Taunton Division had been trying for some time to establish an apiary. In November 1946 a site in Vivary Park was considered. No record can be found as to whether this went ahead or not, but in February 1947 the committee agreed that there would be two hives in the apiary and that two members would be in attendance on two nights during each week!

In February 1954, the committee discussed the possibility of establishing a divisional apiary at Wilton House. A committee of Messrs Hender, Marston and Sparks was given permission to spend up to £15. Mr S. A. Bradbury, proprietor of the garden and seed shop in North Street, Taunton, offered to give a National hive, Dr Harding a W.B.C. hive and a Mr Ready a nucleus in May. Mr Bradbury and Mrs Hender were also to approach John Spiller of Spiller & Webber for help by way of equipment. The committee was obviously very keen. Alas, it was reported to the June committee meeting that the Town Clerk had rejected the request. The next approach was to see if there was a suitable school in the area. Mrs Hender also agreed to approach Mrs Spiller to ask if she would let the Division take over her apiary. The outcome of these approaches is not reported in the succeeding minutes.

The Division's good fortune regarding the offer from Mr Harris was relayed to members ten days later at the AGM on 27th November 1954. In the new year, a committee was to look into the site to see if it was, *"suitable for divisional demonstrations and at a later date, be extended to include instrumental insemination."*

In April, a special committee meeting was called. It comprised of Messrs Vaughan, Hartnell, Styles, Sparks, Bayliffe, Mrs Hender and Mrs Reid. Mr Marston was in the chair. Mr Rolt, the county beekeeping instructor, and Mr Harris were co-opted onto the committee and were in attendance. The subject of a divisional apiary was introduced by the Chairman who stated that through the generosity of Mr Harris, a large hut and a piece of land suitable for beehives would be made available to the Taunton Division.

Mr Rolt considered *"It was a very suitable place and the Taunton Division was very fortunate in having such an opportunity. The first objective would be to get the apiary going".* The sub-committee endorsed the site's suitability for demonstrations. It was suggested that a body of twelve would be needed to run the apiary, and that they should start with three stocks. Mr Harris said the objective was, *"To breed queens of quality and later by artificial insemination".*

On 1st May 1955, just six months after the offer was made, the first demonstration took place at the Taunton Division Heatherton Park apiary. The Chairman placed on record the generosity of Mr Harris in providing the Division with such a splendid site and facilities. Mr Hender had provided a nucleus, a Mr Stone some bees, Mr Styles a Modified National hive, Mr Bradbury and Mrs Harding each a WBC hive. The rest of the season must have been successful as it was later noted in the minutes that there were seven 14lb tins, (three of which were later sold at 2/6d lb), stored in the hut.

On 12th October, Leslie Hender gave the committee an update on the condition of each colony in regard to their queens. He also went on to say that the stinging of members of the public by bees had caused some concern and the secretary was asked to write to Mr Betts, the county secretary regarding liability. A month later, on 11th November, the

The early Post-war Years 57

Division was informed that it was covered for such an eventuality. At the last meeting of this memorable year, the AGM on 26th November 1955, Mr Harris was made a vice-president of the Division and Leslie Hender was profusely thanked for all the hard work he had put in to set up the apiary.

There was concern the following January when some windows of the hut were damaged and entry gained. Some of the hives had also been damaged by gales. The committee was also concerned that legal transfer of ownership of the site was being delayed. By April repairs had been made but the bees were found to have AFB. There was no compensation forthcoming from BDI as the division had failed to pay a separate insurance premium for the apiary. In June their luck changed. The committee agreed the Trust Deed and Mrs Bradbury clarified the misunderstanding with the Somerset Association, that her late husband's equipment was to go to the Taunton Division for its new apiary and not to the county body.

By the end of the year, the deeds had been completed, the boundaries clarified and the area fenced. Pat Yea offered to install electricity free of labour charges. At the AGM on 1st December 1956, Mr Harris was made an Honorary Life Member and Leslie and Lillian Hender were presented with a folding flap tea trolley in appreciation of their work in connection with the setting up and running of the Heatherton Park apiary.

Today, Heatherton Park apiary remains the most important asset owned by the Taunton Division. For fifty years it has provided a meeting place for members where educational demonstrations and lectures continue for their benefit. The occasion was celebrated on May 8th 2005, when the BBKA Chairman, Ivor Davis, was the special guest. Also present was Pat Yea, the only surviving member involved in its inception in 1955. Hopefully, as plans for expansion and renewal proceed, it will continue to serve members for many years to come. After all, they have yet to do artificial insemination!

Membership drops but School Beekeeping thrives

The decline in membership continued for the rest of the decade. It should be appreciated that the years of post-war austerity were behind us at last. Prime Minister, Anthony Eden, remembered for his misjudgement over Suez, resigned through ill health, to be succeeded by Harold Macmillan – Supermac! It was he who made the proud boast that *"most of our people have never had it so good".*

An increasing number of people became the proud owners of motor cars and television sets, worked long hours, often doing overtime, and enjoyed the weekends with their families, cars or in improving their properties. In general people were enjoying a higher standard of living. Beekeeping, like growing one's own vegetables, became a less attractive occupation for those leisure hours. Throughout these years the Association, although becoming smaller, held on to the keener members and remained very active. Falling membership meant that income from subscriptions never matched that which was anticipated when budgeting, and from 1957, for some years a slimmer and less costly year book was produced. Rallies were held to such places as Buckfast Abbey and a weekend school was an annual event at Dillington House where the very best speakers were in attendance.

By the end of the decade membership had levelled off at about 800; little more than half of what it had been ten years earlier. Against this decline in membership it must be said that in colleges and schools beekeeping was an important and thriving, mainly extra-curricular activity. A great number of students at Cannington farm Institute and at Newton Park College, Bath, studied beekeeping and took the Preliminary (now the Basic) examination of the BBKA through the 1950s and 1960s. Many secondary schools in the county are listed as members. Large numbers of pupils took the Junior examination of the BBKA and a few attempted the Preliminary. According to the records the most active schools were: Bedminster Down Secondary Modern, Bridgwater – Sydenham Secondary Modern, Bruton High, Crewkerne Secondary Modern, Highbridge – King Alfred Secondary Modern, Midsomer Norton – Somervale Secondary Modern, Minehead Secondary Modern, Stanchester Secondary Modern, Taunton – Ladymead and Priorswood Secondary Modern, Wellington Secondary Modern, West Monkton Secondary Modern, Wincanton – King Arthur's Secondary Modern, Wiveliscombe – Kingsmead Secondary Modern, Writhlington Secondary Modern and Yeovil High School. Mr Rolt gave tremendous support to school

Girls of Yeovil High School Beekeepers' Club, 1954. Photo: H.C. Tilzey

The early Post-war Years

beekeeping as part of his job. Ultimately a school beekeeping club depended on having a competent beekeeper on the school staff. It would seem, therefore, that to be a beekeeper was almost as important an asset as being a pianist when it came to applying for a teaching appointment, especially where Rural Science was taught. Many members gained their first interest or experience of beekeeping through being introduced to the craft by a teacher at school.

At Bedminster Down School for instance, Mr Desmond Wright was in charge of Rural Science and ran the beekeeping club in a way that benefited the pupils involved quite broadly. The club started up in 1958. Pupil, J. Stabins was elected as Secretary and Treasurer. The club had a space on the school notice board and it met twice a week. The subscription was in the form of shares at two shillings (10p) each. There was a rule limiting the number of shares a person could hold. There were 23 shareholders the first year, including the headmaster. Pupil J. Stabins wrote in the school magazine *"This has been quite an exciting and educational year for the members. Although we have had some sprints away from the hives, even by the experienced beekeepers, the members, I think, will the next time, be more confident in handling bees. We also hope that sometime next year some of the members may be able to take an examination through the Somerset Beekeepers' Association. Because of the bad summer the bees have not been able to gather much honey, but fourteen pounds were obtained from one hive,*

Members at the home apiary, Buckfast Abbey, 1958.   *Photo courtesy of Mrs R. E. Sharratt*

fetching in £2. 5s. 3d (£2.26). The honey jars were exhibited in the entrance hall on the morning of the last day of term. By afternoon school the honey was all sold."

In the school magazine for December 1961 R. Sayer of class 4F wrote: *"This has been a good year for beekeepers and we extracted about one hundred pounds of honey from our hives. We also divided one colony to form two nuclei, so increasing our colonies to four. This year the following boys gained the Junior Certificate of the British Beekeepers' Association: P. Jackson, R. Sayer, W. Willcox and A. Williams.*

This year for the first time we staged a beekeeping display at the Bristol Horticultural Show which attracted a good deal of attention and at least one of our members came away hoarse from answering questions which came at him thick and fast for three hours…"

The summer rallies continued to be a highlight of the Association's calendar and in 1958 was again at Buckfast Abbey. Unfortunately the weather was wet, evidenced in the accompanying photographs, but this did not prevent a strong gathering from enjoying the day. Members were shown the home apiary and the honey-producing plant. They were then taken on to the moors to the queen-mating station where Brother Adam and Father Leo opened nucleus hives and explained how their system worked.

Brother Adam talking to members at the mating station in 1958. Mr. F. J. Pullen, Hon. Sec. of Shepton Mallet Branch, on left.
Photo courtesy of Mrs R. E. Sharratt

CHAPTER 6:
The 1960s and 1970s

1961 and Honey by the Bucketful

Although the prospects looked poor until June and many colonies became short of stores, there was a heartening upturn towards the end of this month. Suddenly nectar was flowing and continued to do so throughout July, ceasing suddenly on 1st August. Very good surpluses of excellent quality were obtained and some beekeepers ran out of supers. At Taunton Show there were 200 entries in the honey classes and nearly 400 at Radstock.

A Winter to remember

In June 1962 David Charles began keeping bees. He started with swarms and by the end of the summer was the proud owner of three colonies, all on new combs and very generously provisioned for the winter. The country narrowly missed a white Christmas that year, but on Boxing Day it snowed heavily, blizzard succeeding blizzard. The southern half of Britain and Somerset in particular suffered greatly. A record published by Mr A. W. T. Hamblin, the Divisional Surveyor at Clutton, near Bristol shows that at Litton, near Chewton Mendip, between 26th December 1962 and 3rd January 1963 snow fell on seven days making a total depth of 705 mm (28.2 inches). Between Boxing Day and 2nd February day temperatures exceeded freezing point on only two occasions and night temperatures fell to a record low level.

The snow became compacted and roads cleared by snowploughs became covered again as winds caused drifting. More snow followed in February. Everybody was affected by the severe conditions, but particularly farmers and the travelling public. Mr Hamblin described the situation as a calamity, the weather being more atrocious than any experienced since well before the 20th century. Many apiaries were beyond reach of their owners and had to take their chance; there was little that could be done by way of assisting the bees to survive. Any bees tempted by sunshine to venture forth became chilled and fell dead on the snow. Hilly areas were the worst affected; the icy spell was so prolonged that snow still lay on the north side of hedges and in ditches on the Mendips, Quantocks and Exmoor in April.

David Charles was living at Windsor at this time. The River Thames was frozen over to the extent that it could be walked and skated upon. Hundreds of lorry loads of compacted snow and ice were cleared from the streets of Windsor and dumped in the Thames. As a beginner, he feared for his bees. In February, on peering through the feedholes, there appeared to be no sign of life within and he wrote them off as killed by the cold. Imagine his joy then, on 3rd

March, when the temperature rose at last and the bees flew in force from all three hives.

The new Hon. Secretary, Mr J. C. Reynolds from Bristol, wrote in his report that 1963 would be remembered as the most difficult and disastrous year in living memory for many beekeepers. Colonel E. C. Brown of Clevedon, responsible for the annual "Crop and Weather Report" published in the year book, claimed that the three months from December 1962 to February 1963 were the three coldest months since 1740, and the most snowy for 150 years. He also stated that half the stocks of bees in Somerset died out during this period. One noted Somerset beekeeper, he says, found a "Wood Clack" *(sic)* right inside one of his hives, but being a sportsman, could not bring himself to wring its neck, which was just as well because there were very few about now.

The President, Mr L. E. Snelgrove, admitted to sustaining heavy colony losses. To assist in replacement he had purchased two packages of Caucasian bees from the USA through Mr Hannam of Birdwood Apiaries, Wells. Using package bees was a popular method of colony increase at this time and Mr Snelgrove gave members detailed directions for hiving them. He remarked that notwithstanding the long journey by land and air the bees arrived in perfect condition.

The summer of 1963 was poor. There was a shower on St Swithin's Day after which there was fine weather until 19th August but the bees seemed to work bramble in preference to the fields carpeted with white clover. The honey season was generally poor, but swarms abounded and there was much replacement of lost stocks. The rally to Buckfast Abbey was marred by rain and a consequent change of plan. The weekend school at Cannington, however, was a highly successful event. In the autumn, King Arthur's School, Wincanton won first prize in the schools section at the National Honey Show in London and was awarded the School Apiary Shield, a perpetual trophy presented by E. H. Taylor, Ltd, the beekeeping appliance manufactures of Welwyn.

Deaths of the President and other prominent Members

1965 was a landmark year and a most sad one in the annals of the Association. November 21st saw the passing of that much loved President, the man who re-established this Association in 1906 and nurtured it over so many years: Mr L. E. Snelgrove. The news was received with profound sorrow for he was highly respected, even revered, and had become a legend in his time. Mr Leslie Hender succeeded him as President, Mr Sawyer became Chairman and the Association's activities continued to run in much the same way as in recent years. At the Bath and West Show the Snelgrove Memorial Shield was competed for by school beekeeping clubs in the South-West counties. The first school to win it was Sydenham Secondary School, Bridgwater.

In the summer of 1966 Dr E. M. Tustin, who was Chairman of the South-Eastern

The 1960s and 1970s

Division collapsed whilst examining a hive of bees and died later in hospital from the effects of a stroke. The BBC and the press, never slow to miss the opportunity of sensationalism, reported that he had been stung to death by bees. This was not the case. The following year, 1967, Mr F. G. Sparks (Fred), who lived at the quarry, Charlton Mackrell, died in August. He was a longstanding member who was briefly the visiting expert for Somerton Branch from its foundation in 1939 until he was called up for war service. After the war he continued this task and also served the Division in other capacities for many years. His father, Mr Tommy Sparks, also a beekeeper and well known member of Taunton Division died within two weeks of his son. Then in December, Col. Prevett, Chairman of the Somerton and District Division died tragically as the result of a car accident. Membership stood at 588 in 1966 and 584 in 1967.

Fred Sparks tending a powerful colony in a WBC hive.
Photo courtesy of D. G. Morris

Tommy Sparks transporting supers by bicycle.
Photo courtesy of Mrs R. E. Sharratt

A new Member for Street and Glastonbury Division

In 1968 David Charles was appointed to a teaching post in Glastonbury and in August came to live there from Buckinghamshire. He took his bees to the New Forest and sited them on the heather with a view to securing a site at Glastonbury and then fetching them to their new home in September. He joined the moribund Street and Glastonbury Division and also joined the Somerton Division as an associate. It was not long after the death of Mr F. R. Underwood that he became Hon. Secretary and Treasurer of Street and Glastonbury

Division and had the satisfaction of building it up. During these years a Wedmore and Cheddar Branch of the Division was re-established. It later became a Division in its own right, and unlike previous groups covering this area has, under sound leadership, endured and prospered.

Loss of full-time County Beekeeping Instructor

In October 1968, after 25 years in the post Mr Arthur Rolt NDB retired from his position as County Beekeeping Instructor. Although regarded as a full-time appointment, Arthur had for several years, been seconded to the Ministry of Agriculture each summer to do advisory work in connection with the Foul Brood Diseases of Bees Order. At a Council meeting held in May of that year it was agreed that the Chairman, Mr R. W. Sawyer, should write to the Chief Education Officer making a strong case for retaining a full-time Instructor after Mr Rolt's retirement, particularly in view of the large number of school clubs operating. At the October meeting it was learnt that in view of necessary economies it was not proposed to appoint a full-time officer in his place. Negotiations followed and Governors of the Somerset Farm Institute were lobbied for their support. A compromise was reached.

In July the Principal of Cannington Farm Institute sent details of a new scheme to meet the needs of Somerset beekeepers. Mr D. A. Husband, NDH, newly appointed Senior County Horticultural Instructor, an experienced beekeeper and a qualified teacher would be available part-time as from 1st August 1969 for beekeeping lectures, demonstrations and advisory work. An adult bee disease diagnosis service would continue, the Institute apiary would be maintained with modifications, but schools work would no longer be on a regular basis. David Husband certainly did his best for the beekeepers of Somerset over many years, but it was in addition to his main work. But for the persistence of the SBKA Council, led by its Chairman, a County beekeeping service would have ceased altogether. The new arrangement was, however, in line with a national trend.

The Hon. Librarian reported that the Bigg-Wither library was not being used and that not a single book had been borrowed during the current year. The probable reasons for this, he surmised, were the very heavy postal costs and the excellent service that is available from local libraries from where any book may be borrowed for a nominal charge. Council agreed to investigate schemes whereby this library could be more readily available to members and students. Mr Snelgrove had recently before he died, donated his entire set of SBKA year books from 1908 to the Bigg-Wither library. It remained little used until 1971 when it was located in the controlled environment of the library on the new campus at the Somerset Farm Institute, Cannington as part of the deal negotiated for beekeeping.

Summary Rally to Usk

The first Severn Bridge, which opened in 1966, made a visit to the beekeeping department at the Monmouthshire Agricultural College at Usk a reality. For many members who attended this visit on a fine day in June 1969 it was their first opportunity to experience crossing the bridge and wondering at this great feat of engineering skill.

The accompanying photograph taken on arrival at the college shows some of the party. The principal characters are, left to right: Mr A. C. Rolt, Somerset CBI, (leaning forward), Mr Amery, Monmouth CBI, Mr L. Hender, President of SBKA, the Principal of the College, Mr C. (Gassy) Harris, Mr T. Sparks, both of Taunton and District Division. Mr A. V. Pavord demonstrated practical methods of manipulating bees in various types of hives, including the new plastic hive.

Summer Rally, 1969. Members meet the College Principal on arrival.   *Photo courtesy of D. G. Morris*

The Price of Honey

In the early 1970s members who sold honey were charging approximately five shillings (5s) a pound to customers at the door. The wholesale price to retail outlets was between 4/6d and 5s a pound. On 15th February 1971 the decimal currency system was introduced and people had to acquaint themselves with new pence. Although the cost of honey remained the same beekeepers had to ask for 25p instead of five shillings. Although they were no worse or better off in their sales, it was noticeable that the price of many commodities seemed to have been rounded upwards in adjusting to the new system.

Formation of the South Western Joint Consultative Council

By the early 1970s dissatisfaction was again being expressed about the inadequacy of the British Beekeepers' Association in fulfilling its role as our parent body. Owen Meyer, the General Secretary at this time accepted an invitation to address the SBKA Council at its meeting held in Bridgwater on 8th January 1972.

At this meeting, chaired by Rex Sawyer, Mr Meyer outlined the work of BBKA, placing it in four categories: administration, services to members, activities and communications. Discussion and questions followed. Mr Meyer's points were summarised fully in the minutes of the meeting, the final paragraph ending with the sentence: *"Mr Meyer was thanked by the Chairman"*. The minutes make no mention of Mr Sawyer's summing up. Thank him, he did, but also sent him packing back to Kent with a flea in his ear. David Charles was witness to events and he well remembers the Chairman's final words: *". . . and that's what we think of your BBKA."* David felt sorry for Owen Meyer. Later he came to know him quite well while serving on the Executive Committee of the BBKA. He was a perfect gentleman, but also man enough to take this rebuff squarely on the chin.

The disquiet continued and a working party was set up to examine the present constitution, rules and management of the BBKA. It consisted of Messrs Rolt, and Jackson, Miss Bindley and the Chairman and Hon. Secretary *ex officio.* Fruitful discussions were held throughout the summer. Council approved of the direction the working party was taking and gave it a unanimous vote of confidence.

In 1973 SBKA convened a meeting of delegates from Somerset and other Associations in the region which was held at the Castle Hotel, Taunton on 24th February. Present were representatives from Devon, Dorset, Gloucestershire, Hampshire and Somerset with apologies from Cornwall, Wiltshire and the Isle of Wight.

Mr Sawyer opened the proceedings by recounting the circumstances that led to the meeting. For some time SBKA had been very dissatisfied by the poor quality of the services offered to members by the BBKA. Nine tenths of its income, it was asserted, was being spent on administration and only one tenth on services to members. Somerset, he said, had appointed a working party whose main objective was to try to find a solution to the problem of this imbalance.

For Somerset, Mr T. Jackson outlined a suggestion that the BBKA could be re-structured to allow regional representation. This entailed the setting up of regional councils and would result in a smaller executive council and the need for holding an annual meeting of delegates in London. There were several objections to this idea. Mr Rolt then listed thirteen items where Somerset would like to see an improvement. It was agreed that consideration be given to the formation of a South Western Counties Consultative Committee and a further meeting was arranged for 2nd October.

The 1960s and 1970s

At this second meeting, the delegates, after consultation with their own Associations, unanimously agreed that this committee should be established. The objective was *"to further the interests of beekeepers by consultation between the county organisations of the south-west"*. The committee would have no executive powers but would be a forum for debate and exchange of ideas. In the absence of Mr Sawyer, Mr Rolt was in the chair, but these two were elected as Chairman and Vice-Chairman, respectively. Gloucestershire and Wiltshire declined to join. (The latter joined in 1979). By 2005 this committee had outlived its predecessor considerably, still meeting regularly twice a year and Somerset taking its turn at providing the chairman. Special mention should be made of Neville Hawken from Bath, at one time a Taunton Division member who was an energetic Hon. Secretary and Treasurer for twelve years.

The County Honey Show

The year 1973 saw the introduction of a honey show that is still held annually. For the first few years it was held at Cannington Farm Institute in conjunction with their Open Day, assuring the show of a good audience. Subsequently it was moved around to be hosted by various Divisions at Burnham-on-Sea, Glastonbury, Huish Episcopi and Shepton Mallet before settling permanently with Taunton and District where it is held each August in conjunction with their own show at the Taunton Flower Show.

A Tree for Bees and People

After the tree-planting at St. Dunstan's School in 1974. Left to right: Mrs Brake, David Brake, Mr and Mrs Cozens, David Charles, Derek Loud, David Loud, June Loud, Michael Miller, Austin Fear, Jack Hepworth.

Photo: C. H. Brake

In 1973 there was a national tree-planting campaign: "Plant a Tree in '73". This was followed by another with the slogan: "Plant some more in '74". On 16th March 1974 the Street and Glastonbury Division played host for the Annual General Meeting. The Division had purchased a fine young specimen of a red-twigged lime tree, *Tilia platyphyllos rubra*. On the morning of the meeting, held at St Dunstan's School, Glastonbury, the tree was planted with due care and ceremony by Mr C. H. Brake, the Divisional President. David Loud, a pupil at the school and a junior member, thanked the Division on behalf of the school. Over thirty years later the tree has grown into a fine specimen laden with blossom in summer and worked avidly by bees. This proved a very rewarding exercise, but how much better a whole avenue would have been!

David Loud in 2005 admires the lime tree he helped to plant.
Photo: A. D. Charles

A new Association is formed

As a result of the Redcliffe-Maud Report, in 1975 a new county of Avon was formed based largely on the Bristol and Bath area. A group of activists in the area felt that advantages were to be gained through breaking away and forming their own Association. At the SBKA Council meeting held in January 1973 the Chairman, Mr R. W. Sawyer, invited the affected Divisions to investigate the implications of the boundary changes and to decide whether they wished to remain in the Somerset BKA. Mr Tom Jackson of Brentry, Hon. Secretary of the Bristol Division agreed to convene a meeting. He reported back in the following October. After lengthy discussion it was agreed that *"This Council recommends that the Divisions and Branches that have members who will be geographically in the County of Avon, should consider firm proposals to take effect from April 1974, for setting up an Avon County Beekeepers' Association which will be part of a new Federation of the Somerset, Gloucestershire and Avon Associations and that Gloucestershire members be asked for their support."*

In January 1974 Mr Jackson reported that it was proposed to set up a County Beekeeping

Association, but that members of the four divisions affected wished to remain as members of the SBKA until at least the end of the year. He asked what SBKA was prepared to do financially to enable the new Association to be set up. It was agreed that any approach on these lines must be made by the appropriate Avon officials once the Association was legally in existence. The new Association was inaugurated at a meeting in Bath on 1st April 1974 without the South Gloucestershire Branch or the bisected Norton-Radstock Division that changed its name to Mendip so that the members could remain with Somerset. The meeting was addressed by Mrs R. E. Clark, NDB, President of the BBKA. The proposed Federation did not materialise.

A year later the Chairman reported that he had sought legal advice on the financial question. It appeared that there was no legal obligation to give a portion of the Somerset BKA assets to Avon, but there was a moral one. It was agreed that a fair share of the assets would be £90, of which £25 had already been handed over as a bridging loan to get the Association started and that this proposition should be put to the next AGM. Avon County BKA was wished every success and hope was expressed for future close co-operation. For 1975 a year book was published jointly. Although 1974 saw a growth in membership from 700 to 783 Somerset anticipated starting 1975 with only 589 members, expecting to lose 194 to the newly formed Association in Avon. In reality, several members from the Bath area preferred to remain in SBKA and the Mendip District Divisional membership increased by 26%.

It was not only a matter of numbers. In effect history shows that this Association sprang from the Bristol group. Somerset lost some very active and experienced members with considerable expertise. Much of its lifeblood was drained away and took some years to recover. In addition, Mrs Elizabeth Lovegrove, the Hon. Secretary and Treasurer for the last five years retired from the post, having married the Vice-Chairman, Arthur Rolt, the retired County Beekeeping Instructor. Both had lost their marital partners and had been on friendly terms for a considerable time. They moved away to Devon. Martin Tovey, recently appointed as General Secretary to the British Beekeepers' Association, took the post. He was young, keen and was doing a good job, but sadly for the Association during his second year he moved to County Durham in connection with his employment. He was succeeded briefly by Mr C. H. Brake of Glastonbury, but by the AGM of 1977 Mr C. A. Harris, also Secretary of the Central Division was elected. While Avon beekeepers forged ahead, the reduced Somerset Association seemed to be somewhat in the doldrums. Membership numbers were maintained, a figure of 691 being reported for 1978.

The main annual events proceeded as usual. The Honey Show was held at Cannington College on Open Day, thus assuring a good attendance and publicity for beekeeping. The arrangement made with Devon BKA in 1960 to alternately run the weekend school between

Above: A meeting at the Gullifords' apiary, Clapton, Ston Easton, July 1977.
Left: Lilian Hender, at the same meeting, shows natural comb built in a skep. *Photos: Terry Hardy*

Seale Hayne Collge and Cannington College continued. The traditional summer rallies still proved popular. They were held at George Vickery's Honey Farm, West Moors, Dorset in 1975, and in 1976 at Sparsholt College of Agriculture near Winchester, which then had a strong beekeeping department. It was to Lackham College near Chippenham with the Wiltshire beekeepers in 1977.

Management and organisational Problems occur

In 1978 there was a very unfortunate misunderstanding over the date of the summer rally to Buckfast Abbey where we were to be welcomed by Brother Adam. It was arranged for Saturday 1st July and three or four Divisions arranged coaches. Many members teamed up to travel together by car and one even altered the date of his wedding to be there! It was one of the best attended rallies ever held, but Brother Adam had been led to expect us the following weekend. He was not there and we were not made welcome. A disgruntled member reported the matter to the local press and the following week the "Bridgwater Mercury" bore the front-page headline *"Bee-keepers told to buzz-off."*

In 1979 there was a delay in publication of the year book containing notice of the annual general meeting. It was available only just before the meeting on 17th March and many members had not received their copy. This was the last straw for the Chairman, Rex Sawyer. He felt angry, embarrassed and let down and promptly resigned because of the un-businesslike way in which the Association was being run. David Charles, as Vice-Chairman, unexpectedly found himself at the helm.

Improved communication with the membership was a priority and the launch of a thrice-yearly newsletter was therefore an important step forward. The newsletter has been produced regularly since this time.

Bee-keepers told to buzz-off

OVER 200 bee-keeping enthusiasts from all parts of Somerset were politely told to "buzz off" on Saturday when they travelled over the border into Devon.

The 250 strong party, including a group from Bridgwater, were members of the Somerset bee-keepers association.

And Saturday was the day of their big rally. The men and women went in cars and coaches from towns throughout Somerset to Buckfast Abbey especially to see Brother Adam, one of the most famous bee-keepers in the country, with his strain of hybrid bees.

WRONG WEEK

But when they reached their destination the bee-keepers were told: "Sorry, you have come the wrong week. We were expecting you all NEXT Saturday."

One of the Bridgwater group said on Monday: "This type of thing could only happen in Somerset. There we were standing around for over an hour all wondering what had gone wrong. We were obviously puzzled because we expected the place to be a hive of activity on our arrival."

He added: "There was obviously a big misunderstanding because we thought the visit was definitely planned for last Saturday.

"All the bookings had been made for that day and coaches laid on for the outing to Buckfast from various parts of the county. The date, July 1, was even in our association diary.

"Unfortunately I don't think we can go this year again. It's just too far to travel and it would be extremely difficult to organise a second time. Anyway we had a good laugh later on. What else could we do but try to see the funny side!"

The dates' muddle will be discussed by bee-keepers at a committee meeting later this week.

Rex Sawyer displays a touch of humour on the ill-fated day in 1978 at Buckfast Abbey.
Photo: A. D. Charles

1979 was also notable for being a very poor year for bees and honey. The weather was either cold, wet or both from the new year until May, with snow in March on several days. Not until September was there a settled spell of good weather. This was also the year when we heard about the threat from a parasitic mite named *Varroa* and read of its steady advance across Europe towards Britain. A second attempt was made to hold the summer rally at Buckfast Abbey and this time it went well. Fewer members attended but a coach of Avon beekeeping friends joined the gathering. To see Brother Adam handle a colony and hear his wise words was a memorable experience. Alas, the press was not interested in a successful event.

Brother Adam explains to members the design of his queen-mating hives and feeders, July 1979.
Photo: K. A. T. Edwards

Chapter 7:
The 1980s: a Decade of Decline

After three years in the post Chris Harris, for personal reasons, resigned as Hon. Secretary at the 1980 annual general meeting. It proved impossible to find a replacement and, as Chairman, David Charles found himself doing much of the secretarial work for the first few months. In September of that year Cmdr John Goodman took up the post. He showed all the qualities one could wish for in a good and efficient secretary. Great regret was therefore expressed when he announced, shortly after being confirmed in post, that he would be unable to continue. His career was taking him to the north of England so that he would be leaving the area (he much later became a Regional Bee Inspector).

At the AGM of 1981 Miss Bindley announced her wish to retire from the Presidency and Mr Rex Sawyer was elected to succeed her. Mr Arnold Downes, a Junior School headmaster in Wells and keen member of the Central Division, although not present, had offered his services as Hon. Secretary. Those present felt that a volunteer was probably better than ten pressed men and duly elected him. David Charles, having had a difficult year without an Hon. Secretary for much of the time, his professional life becoming more demanding and his personal circumstances having changed, felt unable to undertake a third year as Chairman, eligible to him under the new rules. He was also becoming heavily involved with the work of the BBKA. His final task as Chairman was to produce the minutes of this AGM because neither the old nor the new secretary was present; neither was the minuting secretary. Such was the position when Ken Edwards, the Vice-Chairman was elected to the chair at this meeting. He too was in for a difficult time.

The May Council meeting held at St. Dunstan's School, Glastonbury was a very unsatisfactory occasion. Despite the Chairman having previously discussed the agenda with the secretary it had not been drawn up and circulated in advance to Council members according to the rules. The Chairman promptly wrote one in chalk and this was displayed on a conveniently placed blackboard. Mr Downes, Hon. Secretary, and Mr Sawyer, President, were absent from this meeting, the latter claiming not to have known about it.

Rex Sawyer, without consulting the Chairman, subsequently instructed the Hon. Secretary to call a meeting of the emergency committee for 9th June. This was an inconvenient date for the Chairman, who, in any case, had not been consulted on the matter. Resulting from this, Mr Edwards not only had a long telephone conversation with the President but set out quite clearly the position in a long letter to him; a letter that the President described as *"somewhat impertinent"*. On 29th June Mr Sawyer circulated a letter to the emergency committee and divisional secretaries headed "The organisation of the SBKA" that ended with his expressed hope for a return to the smooth and democratic running of the Association. Meanwhile, the Chairman was applying pragmatism to the situation and endeavouring to move things on. Together with David Husband they had made all the arrangements for the weekend school at Cannington. To the ordinary member all appeared to be proceeding as normal.

The October Council meeting, held at Cannington, was attended by all the officers except the President. The minutes of the previous meeting were considered to be unsatisfactory owing to inaccuracies and omissions. It was resolved that they be re-drafted and presented again at the next meeting. Apart from that it was business as usual.

Council met again in November and in the absence of Mr Edwards, was conducted by the Vice-Chairman, Mr M. Priscott. It was at this meeting that the secretary stated he had received the grant from the EEC Intervention Board. The money was distributed to divisions according to the returns made at the rate of 62 pence per colony. The total amount received was £1,864 to be spent as dictated by the rules of the Intervention Board.

At the February 1982 meeting all seemed to be going well and the minutes, previously rejected and now amended were approved. But all was still not well between the senior officers. At the Council meeting of 9th May the President raised a point of order. He stated, after seeking legal advice on the matter, that this meeting had not been properly convened and he asked the Chairman to declare it out of order. Furthermore, he indicated that he would *"take the chairman to law"* if he allowed the meeting to proceed. Mr Edwards then asked those present to consider whether they would continue to accept him as Chairman of the meeting if he were to declare it in order. A vote was taken, the majority declaring in favour of the Chairman. The Chairman declined to declare the meeting out of order and at this point the President left the meeting. Normal business resumed and no litigation ensued.

The rest of the year went quite smoothly, with events, the covenant scheme and a committee at work on the proposed amended rules. The honey show was held at Bridgwater Town Hall but the summer rally to Cardiff University was cancelled owing to lack of support. This appeared to be because of the high cost of the transport and a requirement to pay £7.50 per head to the University.

Matters generally proceeded better in 1983. The summer rally was to the College at Pershore in Worcestershire, and the weekend school at Cannington was held in conjunction with the British Isles Bee Breeders' Association. Few of our members attended, but the event was a financial success. New rules were drafted, but Somerton and District Division complained about not being represented on the Rules committee. The Division called an extraordinary meeting regarding the proposed new rules and it met on 25th August. The Council meeting held at Cannington on 9th September was ruled out of order. There is no

The 1980s: A Decade of Decline

record of the attendance, but presumably there was not a quorum, and again the Hon. Secretary was absent. No agenda had been circulated by post in advance but those members present voted to continue with the meeting for discussion and exchange of information, with the agenda again written on a blackboard. At the November meeting it was agreed that the meeting was out of order and that no decisions were taken. With the exception of Mr Sawyer members voted to accept the minutes such as they were.

On 26th November Council met in Glastonbury when the attendance was much higher. There is no mention of the Hon. Secretary in the minutes but a replacement was nominated at this meeting. This was Mrs Esther Farnes of Huish Episcopi. It was noted that several divisions had received a visit from the Intervention Board Officer who was checking on how the grant was being spent, the purposes for which had been laid down. There would be several changes of officers at the next AGM, and nominations were still required for the chairmanship, as Mr Priscott, the current Vice-Chairman did not wish to stand. It was not surprising under the circumstances that no other members were willing to allow their names to go forward.

At the AGM held on 17th March 1984 at Yeovil the Chairman, Ken Edwards, having completed three years was not eligible for re-election. As the Hon. Secretary was again absent and was not continuing in the post, the Chairman reported on the year's work. He referred not only to our successes but also to the difficulties that there had been and he made a plea for all officers to be treated with consideration. *"Without officers"*, he said, *"the association would cease to exist. The work involved entails a great deal of time and effort. All this is done without expectation of financial reward and it is most unreasonable then to make their work more difficult by giving them a hard time".* He paid tribute to the work and wise counsel of Jack Hepworth who had been an excellent treasurer and to Michael Duffin who had spent so much time and effort in trying to establish a covenanting scheme. No nomination had been received to fill the chair but Mike Duffin, who had accepted nomination as Vice-Chairman was duly elected to this post knowing that he would, in all probability, be straight in at the deep end. Mr and Mrs Farnes were elected to the posts of Hon. Secretary and Treasurer respectively. Both Ken Edwards and Jack Hepworth were elected as Vice-Presidents in appreciation of their work for the association.

The May 1984 Council meeting was held at St Dunstan's School, Glastonbury with Mike Duffin in the chair. His first task was to report the death of our newly appointed auditor, Mr Alan Bromley, former Somerton Divisional Treasurer, who had become seriously ill and died the previous month. No one else having come forward, Mike Duffin agreed to become Chairman and Mr F. J. Horne of Yeovil was elected Vice-Chairman. Mr and Mrs Farnes agreed to continue in office, if Council wished, until their move from the area. Thus an impending crisis having been averted, the work of the Association continued. However,

Mr J. Fieldhouse informed Council that he could not continue as year book editor after the 1985 edition. These difficulties in finding suitable members to hold office and remain in office dogged the 1980s. The remaining years of the decade passed rather uneventfully, Council endeavouring to keep the ship afloat rather than making much progress.

Rex Sawyer worked tirelessly for the good of the Association for over thirty years. He cared for it greatly and his actions were from serious concern, even fear that it might slide down a slippery slope towards demise. He was a man of precision; he felt strongly that procedures and rules were there for a purpose and should be followed and obeyed. His interventions were well intended for the good of the Association and readers should not misjudge him through his actions in the recent difficulties.

A View of Secretaries and their Role

This seems a convenient juncture at which to consider the views of the late Herbert Mace, who figured briefly in an earlier chapter. In 1928, subsequent to a survey he had made, Mace wrote a booklet entitled "Beekeepers' Associations, a critical Survey." With regard to the role of secretary he felt he was *"treading where angels had feared to tread,"* but these are some of his main points: *"I take off my hat to every man or woman who has the pluck to undertake what is often a most thankless task. I know full well how exasperating it can be, for in various spheres of life I have acted for long periods as an hon. sec. Those who take on the job, either from personal desire, at the wish of members, or even because no one else can be found to do the work, take it with all its responsibilities, and it is far better for the Association and the man himself that he should refuse the office rather than let the Association down by doing the work indifferently.*

A secretary should always bear in mind the fact that he is only the earpiece and mouthpiece of the Association, not the sole director of it. It behoves every Association to use great care in the selection of a secretary. A common pitfall is to choose the man who knows most about beekeeping, sometimes a fatal error, for the qualities which make the successful beekeeper do not always go hand-in-hand with secretarial ability. Indeed, it is only too apparent that many secretaries have the crudest notions of the work, their letters being on the verge of illiteracy.

If the Association is to prosper it must be always progressing, and the secretary is necessarily the one to lead the way. To sum up, I would submit that the best qualifications of a secretary are tact, judgment and method, and with these three must, of course be a blend of enthusiasm for the cause. It is a hard life, my master, but it has its compensations, and when all is said and done, virtue is its own reward."

Times and circumstances have changed since these views were expressed. Some of what Mace says is still applicable; certainly his views contain food for thought. These days the social climate is such that very few people are able or willing to put themselves forward for office, and this is the first stumbling block when making an appointment. There are either no applicants or no choice of candidate for the post. This situation is now compensated for to some extent because the work has diversified, responsibilities and accountability being greater. Consequently, management is becoming much more of a team effort and the workload on the secretary has eased. Long gone are the days of clergy incumbents in small parishes and with time on their hands!

Secretarial Post vacant again

Another setback was the retirement of David Husband as part time County Beekeeping Adviser. It was learned that there were no plans to appoint his successor on the same basis and that beekeepers would lose this valuable support. Mr and Mrs Farnes moved away, as expected and Mr Duffin having completed three years as Chairman took on the treasurership *pro tem*. He actually served a three-year term before he too left the county for Hampshire in connection with his work in the banking profession. For a time there was again no secretary. Finding a replacement proved difficult, but through the good offices of Mr J. Gilling of Rooksbridge it was learned that Mrs Margaret Pusey of Sutton Mallet would consider taking it on. There was reluctance at appointing somebody who was neither a beekeeper nor acquainted with the beekeeping scene, but needs must. David Charles and David Husband interviewed her in her home. They explained to her what was involved and recommended her appointment to Council. Margaret bravely took it on, finding the experience a very steep learning curve but did as good a job as one could expect for nearly two years. Regret was expressed when she resigned. Margaret was succeeded by another non-beekeeper, Mrs Barbara Quartly, from Bradford on Tone who proved a very efficient secretary, remaining in the post for eight years.

For some years the summer rally had proved less popular than formerly and numbers attending had dwindled. In 1987 Coleg Howell Harris, Brecon was the venue. Here Charles Dublon, then the Principality Beekeeping Adviser for Wales gave an excellent demonstration on queen-raising. The event was poorly attended and Council decided not to hold rallies beyond the county boundary in future. A month later our biennial summer school, alternating with Devon BKAs at Seale Hayne College, was to have been held at Cannington. Vince Cook, the National Beekeeping Adviser, Dr Harry Riches, Roy Page, Bryan Welch, NDB and local experts who included the wax queen – Elizabeth Duffin, made up a comprehensive and varied programme. Alas, but a handful of people enrolled and the event had to be cancelled at a late hour. This was the last weekend school to be run by the Association.

In 1988 the summer rally was held at Cannington College and considerable trouble was taken in laying on an interesting programme of activities that included wax flower, candle

Cup Winners at the Somerton and District Division's Annual Honey Show and Craft Fayre, Huish Episcopi School, 17th September 1988. Left to right: Amanda Chuter, Laurel Morris, C. H. G. (Sam) Langford, Joyce Bromley, Monica Haynes. *Photo: F. J. Horne*

and foundation making, microscope work on pollen and bee diseases, practical beekeeping, microwave cooking and much more. Dr Mick Street from Bicton College, Devon, gave an illustrated talk on "Bees and Flowers". This event was a success.

Not for the first time in the Association's history, production of the year book proved a problem. The cost of production had soared, advertising revenue was greatly reduced, difficulty was experienced in finding editors and the number of errors in recent editions rendered it unreliable. Even the Chairman's report was signed *"M. Horner"* instead of *"F. J. Horne"*. Much of the content was regarded as irrelevant to most of the members. On the recommendation of a small working party, the SBKA Council decided to discontinue this historic reference book, but instead to produce computer generated membership lists and publish a newsletter at more frequent intervals.

Generally speaking, these were bad years for colony losses and honey production. Good years were the exception. Both Richard Clark of Street and David Morris of Halse have reliable figures showing that 1983 and 1989 were exceptionally good. In the years 1985 and 1986 the honey harvests were particularly low, with heavy colony losses during the latter. The remaining years were either poor or about the overall average.

Between the years 1980 and 1989 membership of the Association fell from 720 to 438. True, there was a national decline, but it was not so severe as in Somerset. The reasons appeared to be that many beekeepers were disheartened by poor results from their efforts

The 1980s: A Decade of Decline

and loss of colonies. In addition, the threat of the varroa mite loomed closer as reports from the continent spoke of its relentless movement westwards. *"Not if, but when it arrives"*, the experts told beekeepers.

To finish this decade on a bright note, after considerable deliberation and due to the persistence of the Chairman, Mr Fred Horne, the part time post of County Beekeeping Adviser was reinstated after a period of nearly two years following the retirement, and sadly, the early death of David Husband. David Charles had recently taken early retirement from teaching in Glastonbury and applied for the post. Two Somerset beekeepers were interviewed and he was the one appointed. At the college he re-established the teaching apiary, organised an adult bee disease diagnosis scheme and ran courses. Peter and Mary Barnes-Gorell were regular attenders.

David Charles, Jack and Peggy Hepworth read a comb. Apiary meeting, Compton Dundon, 1984.
Photo: P. R. McArdle

The re-established apiary in the bee garden, Cannington College, 1990. *Photo: A. D. Charles*

Beekeeping demonstration, Cannington Open Day, July 1990. *Photo: A. D. Charles*

CHAPTER 8:
Towards the Millennium

The Year Book makes a Comeback

Ceasing production of the year book, despite its shortcomings, was an unpopular decision with many members; behind the scenes a small group was looking into the possibility of its re-introduction. At a Council meeting held at Cannington College on 8th September 1990 Mr Brian Kilner of Taunton Division who was engaged in the printing trade, was invited to attend and address Council about modern methods of production. A sub-committee was then formed to investigate the feasibility of re-introducing a year book for 1991, paid for by advertising revenue.

By the following Council meeting in November the sub-committee had met and furthermore, had acquired advertising revenue to the value of £200. This was probably less than half the required amount. A motion allowing the sub-committee to continue its work was not forthcoming and it was agreed that if a replacement editor could be found for the newsletter this should continue for another year. Mr Trood was asked to wind up the project to produce a year book and to return the advertising money already received. The matter was not allowed to rest there, the sub-committee meeting again despite what had been agreed. They discussed producing a reduced format of 48 pages.

The Chairman was determined to have the year book reinstated and a special Council meeting was held in the committee room at the Somerset County Cricket Ground, Taunton, on 5th January 1991, with Brian Kilner present. The main business was to discuss further the possible re-introduction of the year book. The Chairman, Mr John Newcombe stated that when speaking to local secretaries concerning the proposed social function he had raised the matter of re-introducing the year book and found that nine were now in favour. Mr Dell, the new Hon. Treasurer reported an improved financial position and Mr Trood stated that he knew of fourteen potential advertisers. There was lengthy discussion on the matter and eventually a proposal to re-introduce it was carried by nine votes to one. A greatly improved book was produced and publication has continued since that time.

A new Era in Colony Management begins

The British Bee Journal of February 1976 carried an article by Kari Koivuleto from Finland, entitled *"Varroa jacobsonii – a new Mite infesting Honeybees in Europe"*. He stated that this mite, originating in the Far East, was spreading relentlessly westwards and was now in Poland. He further stated that infestation by these mites could destroy a colony of bees in a period of 4 – 5 years. (The species was later found to be *V. destructor*).

By 1982 the mites had been found in Italy, France and Holland. In this country experts spoke of when and not if the mite found its way across the English Channel and into the British Isles. MAFF, local associations and County Beekeeping Officers conveyed information to beekeepers about the impending problem. Beekeepers were bombarded with pamphlets and magazine articles on the mite, methods of detection and control. Demonstrations were given on various techniques to be employed such as the tobacco smoke detection test and how to examine hive debris for dead mites. Many beekeepers did not trouble to look for mites and buried their heads in the sand with excuses such as *"My bees are not near anybody else's".*

Unlike the situation existing earlier in the century when the so-called Isle of Wight disease was wiping out colonies wholesale and was referred to as a mysterious scourge, this time beekeepers would know exactly what to look for and what was the probable cause when colonies began to collapse.

Beekeeping alongside Varroa Mites

Sixteen years after that first alert, on 4th April 1992, Devon BKA, Torbay Branch members found mites during a routine test at their Cockington apiary. By November of that year, MAFF had identified infestations in 200 apiaries from Cornwall to Lincolnshire. In Somerset, of 218 colonies inspected, mites were present in forty of them. Fortunately for beekeepers a synthetic pyrethroid acaricide called flumethrin, with the proprietary name of Bayvarol, was licensed for use in the United Kingdom during that year. This product was costly and many beekeepers failed to check for presence of the mite or to treat it. Colonies were collapsing more quickly than had been predicted.

The Association did all possible to encourage beekeepers to become members so that they could be kept informed, and those not prepared to adapt their methods or improve their husbandry were advised to discontinue beekeeping for the benefit of others. Some beekeepers heeded the advice to join, but others gave up when their colonies died out and the hives became heavily infested with Greater Wax Moth. Most feral colonies succumbed and those that re-established were generally short-lived. During 1993 varroa mites, identified initially in south Somerset, appeared to have spread throughout the whole county. Members were advised *"If you haven't found them by now, you have not looked hard enough."* Re-infestation was causing great concern. Beekeepers learnt to cope with the new situation and during the decade monitoring and treatments became routine. There were several good seasons and good yields of honey were obtained from well-managed colonies. Members came and went, but membership showed only a small decline throughout the decade.

Simon Jones (right) checks a sample of floor debris for the presence of mites during a Varroa workshop run by David Charles at Cannington College.
Photo: A. D. Charles

SBKA Charity Status

As with many beekeeping Associations in the country, Somerset concluded that the advantages to be gained by becoming a registered charity outweighed any disadvantages. After a year of debate it was agreed to proceed with an application and this was duly submitted to the Charity Commission. In April 1980, under the Charities Act, 1960, we were entered in the Central Register of Charities.

The chief benefit of being a registered charity in law was that the Association was able to reclaim income tax paid by members on the amount of their subscriptions when paid under deed of covenant. This scheme had potential for substantially increasing the Association's income on an annual basis. The Treasurer, Mr J. M. Duffin went to considerable trouble explaining to members how the scheme worked and encouraging those eligible to sign the necessary form. It proved difficult to find anyone willing to take on the work and responsibility of operating the scheme and many members were reluctant to participate. No benefit was received from Mr Duffin's efforts until after he had left the district, but when payment was eventually received from the Inland Revenue it was backdated.

In 1990, when Mr R. P. Dell became Hon. Treasurer a serious effort was again made to reclaim tax from the Department of Inland Revenue. Implementing this procedure took a considerable amount of time for Mr Dell because, regrettably, the Association had not kept the Charity Commission acquainted with rule changes or submitted annual statements of account as required. In addition, a query made by the Commission in December 1983 had not been clarified, replied to or acted upon in any way. Mr Dell held protracted correspondence with the Charity Commission and the Department of Inland Revenue to correct the situation. A further amendment to the rules was needed to rectify the matter raised by the Charity Commission in 1983. An Extraordinary General Meeting was called for 30th November 1991 at Glastonbury for the purpose of approving the revised rules of the Association.

Early in 1992 confirmation of the Association's status as a charity was received from the Charity Commission. In 1992 the sum of £567.25 was refunded for the years 1988 to 1991 and was distributed to the nine divisions that had participated. The Association has continued to benefit financially from tax rebates, first under the covenant scheme and more recently through "Gift Aid". In 2004 almost £1,600 was reclaimed and distributed in proportion to the Divisions. This additional income has been of considerable help in augmenting our funds and placing the Association on a firm financial footing. The additional administrative work of the Treasurer and the various divisional treasurers or membership secretaries must be acknowledged, but in particular mention must be made of the President, Mary Barnes-Gorell who has been Covenants and then Gift Aid Secretary for over ten years and has borne the brunt of the administrative work involved.

A Government Innovation and Expert Support arrives

In 1994 the Government announced details of *"a new powerful back-up team for England's beekeepers to help them maintain and sustain a healthy horticulture industry"*. After meeting beekeepers' representatives in London on 10th March 1994, Mr Michael Jack, MAFF Minister of State declared *"I am determined to ensure that our beekeepers receive more help in the fight to ensure that England's bee population remains healthy and able to fulfil its essential role. Beekeeping is of great importance for pollination of crops and wild flowers and for the rural economy as well as the balance of trade through honey production. A healthy bee population makes for a healthy horticulture."*

The main features of this new policy were the appointment of a team of nine Regional Bee Inspectors and a new programme of training drawn up by these inspectors in conjunction with local and county beekeeping groups. Somerset fell into the sprawling Severn Region and Len Dixon was the first RBI to cover this county. He came from Surrey, but unfortunately for Somerset beekeepers he set up home near Leominster, quite some distance away from this county. The RBIs appeared to work long hours with great energy and enthusiasm. In 1999 Somerset was transferred to the South-Western Region. Richard Ball became the new RBI for this Region and came to live near Sidmouth. His services have been greatly used by the Association at all levels, for lectures, demonstrations, bee disease safaris and so on, and his involvement has had a substantial beneficial effect on beekeeping standards in Somerset.

It seems that with improved husbandry and some favourable seasons, during the 1990s average honey yields increased and many beekeepers came to realise that there is life beyond varroosis! Beekeepers were encouraged to use "Integrated Pest Management" rather than relying on chemical methods of control because of the risk of pyrethroid resistant strains of the mite evolving.

Council makes new Appointments in Specialist Areas

Strangely, the arrival of *V. destructor* helped the Association to progress. It had to be well organised and active to cope with the new situation, helping beekeepers to continue successfully in the craft. At the Council meeting held on 27th November 1993 the structure of Council was questioned. It was said at this meeting that the Association lacked force and needed to be more effective. One of two delegates representing the Somerton and District Division, Gerald Fisher, raised the matter. Gerald, and his wife, Stella, of Huish Episcopi, had become members in 1974 and almost from the outset both had been active and held various offices at Divisional level. Gerald had recently retired and felt able to put even more effort into Association work and his attention turned to the central body. Discussion ensued on the matter raised and all delegates were asked to come to the next meeting with positive suggestions on how management should change to make the Association more effective. In the meantime the Somerton Division organised some discussion groups and armed with new ideas Gerald Fisher was invited to address the Council meeting held on 5th February 1994. He said that the group advocated a two-pronged approach, more services to members and more promotion of the Association and the craft.

In the first category education needed priority, with co-ordination of study groups, identifying suitable persons as lecturers and demonstrators, someone to deal with adult bee samples, organising visits to national events, bulk purchases and a pollination service to farmers. In order to promote beekeeping there should be a presence at all major local shows with beekeeping demonstrations and constant publicity through the press. Records show that many of these were not new ideas. Some had been practised in the past but for some reason or another had lapsed. These ideas were no less worthy for that. To achieve the aims set out four new appointments and members to fill them were proposed:

Public Relations: T. Trood
Events: N. Trood

Pollination: D. Bates

Promotions: G. Fisher

These new appointments were approved on a year's trial but have continued to function ever since, their contribution to the lifeblood of the Association having been considerable and two of the officers, Gerald Fisher and Neil Trood, still hold their positions. Not every area has been taken up. For instance, there is no bulk purchase scheme for the benefit of members. What exists, in most Divisions however, is the facility to hire equipment, especially for extracting honey. This is of great benefit to newer members because of the great capital outlay needed.

The End of County Beekeeping Advisers

Another matter reported at this February 1994 meeting was that David Charles had resigned from his part-time post of Beekeeping Adviser at Cannington College The Chairman felt that this was a great loss but blamed it on lack of support for courses on beekeeping held there, partly because not many people were able to attend in the daytime. Michael Milton boasted an enrolment of 22 persons for the Somerton Division beginners' classes held in the evenings. There *were* well-attended evening courses held at beginners' and advanced levels all through the winter evenings at Cannington College also but the big difference was that Somerton Division's courses were free and the College fees were fairly hefty. The Chairman stated that he would contact the College authorities to establish whether it was intended to make another appointment, and meanwhile no adult bee samples should be sent to Cannington. The minutes do not say, that Cannington College ordered the discontinuation of the adult bee disease service and cut David Charles's hours by one hundred per annum, making it barely worth the journey of 22 miles each way from home, depriving him of a rewarding part of his work and beekeepers of a valuable service. The college had, in line with Government policy, become independent and self-financing. There was tight financial control and David felt that the end of the diagnostic service was the thin end of the wedge. Having never been dismissed from a post in his life he decided to pre-empt matters!

The College Principal replied to the Chairman's letter saying that if the County Council was prepared to provide funds an appointee could be based at the college. David Charles looked after the bees voluntarily throughout that summer. Between visits an unexpected incident occurred. An apiary sited behind a large hawthorn hedge near the pig unit on the Hinckley Road was demolished by a JCB employed to grub out the hedge. This hedge was between the bees and the pavement and its removal meant passing schoolchildren walked within feet of three upturned hives, two of which had broken apart with combs toppling out. It was an amazing and unbelievable thing to happen. David was called to the rescue and spent several hours making the hives secure for removal. Some of the honey was removed, the hives loaded on to the college lorry in the evening and taken to a field near Chilton Polden out of harm's way. David's abiding memory is of driving behind the lorry along the Broadway through Bridgwater watching bees being blown away among the passing traffic! It had been impossible to close two of the entrances as bees were clustered upon the front of the hives after their ordeal.

After this incident David made a fair offer for all the hives and equipment owned by the college. His offer was declined; as the storage room was required for another purpose the equipment was dispersed around the college, most of it never to be seen again. The hives were removed from the historic old bee garden to an apple orchard, where the bees subsequently died out.

This was an igniminious end for a small department that in earlier days had been so highly regarded and which was of such beneficial service to the craft. No further appointment was made, but the college would in a few years time, play a very different role in the progress of the craft and the Association.

A Survey of winter Losses

By 1995 the dire effects of the varroa mite were becoming evident and there were many colony losses, especially during the winter of 1995 – 96. Members were encouraged to monitor their bees regularly for the presence of mites so that treatment could be applied. It was found that a colony appearing to be healthy and productive could collapse within weeks possibly due to migration of mites from other collapsing colonies.

It was decided to conduct a survey of beekeepers in Somerset, whether or not they were members.

These are the results:

	Number of beekeepers	Number of colonies 30.09.95	Number of colonies 01.04.96	losses	% losses
Members	103	681	491	190	27.9
Non-members	15	66	10	56	84.8
Combined	118	747	501	246	32.9

Many colonies died leaving large quantities of unconsumed winter stores. It was known, through hearsay that some members who suffered severe losses did not send in a return, perhaps being too lethargic or too embarrassed to do so. Even so, the figures, based on a relatively small number of beekeepers do indicate that the knowledge gained by being a member was advantageous.

An Attempt to re-organise Divisional Structure

When the Wedmore and Cheddar Branch of the Street and Glastonbury Division was granted divisional status in 1985 it brought the total number of divisions of the Association to thirteen and no branches remained. This was a far cry from the five divisions of a then larger county established under the rules adopted in 1920. Many new divisions had been created since that time, generally through branches attaining divisional status. Many branches came and went, particularly during and immediately after World War II. The message sent by this trend was that the three-tier system of central body, division and branch was not popular or practical at local level.

By 1996 Council felt that having thirteen divisions, was unwieldy and expensive in administration now that membership was lower and the county smaller than prior to 1974. Ways of reducing this number were considered. It was proposed that by amalgamation the number of divisions should be reduced to six. Small divisions would not tolerate having amalgamation imposed upon them and stalemate was reached. Ultimately it was agreed that the best way of resolving the situation was to help these small divisions. It was sad, however that two of the oldest groups that in times past were associated with some of our most illustrious members ceased to exist. The moribund Street and Glastonbury Division amalgamated with the Somerton and District Division and the Bridgwater and District Division wound up altogether because nobody was prepared to take office. The number of Divisions was thus reduced to eleven.

Instructional Activities

The tradition of weekend schools and summer rallies became part of our past, members having voted with their feet regarding these events. In their place the Association embarked on holding annual days of instruction, variously known as day schools, study days or lecture days. Initially they were held in the Neroche Hall, a building converted from a redundant church near Bickenhall, several miles south of Taunton. The Association was able to call on speakers of national repute and these occasions were well attended. After some years the event was moved to the Westex Theatre at the Royal Bath and West Showground where there was more capacity and better facilities. The Education Officer, Mrs Caroline Butter still organises these events.

As more became known about the varroa mite and its effects in this country, new methods of monitoring, management and control were devised. The use of formic acid is an example, but precision in carrying out the treatment was required. There were frequently short courses in various parts of the county to assist members in applying new techniques to keep their colonies healthy.

Promotion is shown to bring Results

The Royal Bath and West Show, the National Amateur Gardening Show and Yesterday's Farming, all at the Royal Bath and West showground near Shepton Mallet, the Somerset Steam Spectacular near Langport and the Taunton Flower Show were the main events at which Gerald Fisher, the Promotions Officer, and a team of helpers, organised stands to publicise and promote beekeeping before the public. Part of the spin-off was greater media publicity. For instance, at the 1998 National Amateur Gardening Show a television interview was recorded by HTV on the Association's stand and screened as part of the gardening programme.

Mary Barnes-Gorell, second left, and Alex Morrice, on duty at the observation hive, National Amateur Gardening Show, 1996.
Photo: G. Fisher

Most newer members have joined the Association through an introduction at these events and subsequent introductory courses. Gerald's plan was that divisions closest to the venues for large events would take responsibility for mounting an exhibition for SBKA, but this has only worked to a limited extent and he and his team have borne the brunt of this work over many years. There have been several local initiatives taken by divisions at smaller events.

Somerton Division Beginners' practical Class, Tengore House, Huish Episcopi, May 2000.   Photo: T. J. Harris

Reliable data compiled after introductory courses were run at Somerton, South-Eastern, Wedmore and Cheddar and Yeovil Divisions showed a clear increase in membership against a decline where courses were not held. Without these promotions and subsequent courses membership would have fallen considerably. During the decade there was a slight decline, but it was small in comparison with the national total which fell from approximately 13,000 to something over 9,000 members of associations affiliated to BBKA.

In 1999 David Morris of Halse took on the post of Publicity Officer and since that time the Association and its work have featured much more prominently in the media.

Pollination Service

Each year a few members had been making their own arrangements to take hives to crops for pollination in return for an agreed fee. It was felt that a co-ordinated scheme should operate which would provide a pollination service for small growers not in a position to use the professional service offered by bee farmers. This scheme would not only assist members and growers but also broaden the Association's areas of activity and demonstrate its awareness of the value of bees as pollinators.

David Bates of Curry Rivel was appointed as Pollination Officer. He set up a scheme for liaison between beekeepers and growers governed by an agreement and he made the arrangements for each contract. The Association collected the fees, apportioned them and distributed them. The fee in 1994 was £14 per hive. Of this the beekeeper received £10, the central body £2 and the beekeeper's Division £2.

In 1996 Veronica Watts of Long Sutton took over as Pollination Officer. During her term of office the service became a little more widely used. In 1999, however, Veronica had reported to Council that although ten members had offered to take hives to crops she had received no enquiries from growers. Following consultation with the BBKA Technical Committee the future of the service was discussed at length at a Council meeting held on 12th February 2000. It was decided to give wider publicity to the scheme and Veronica then produced the Association's leaflet on its pollination service, directed towards both members and growers. It made the case for crop pollination and spelled out the responsibilities of both beekeeper and grower when a contract is being fulfilled. There was a great improvement in 2000, it being the most successful year to date. For the first time there were contracts for the pollination of cider apple orchards. The apportionment of the fee having been altered, receipts from the growers, who all paid up promptly, generated a total of £52 for the Association and £338 for the members who participated.

Thus, the century, and indeed the millennium ends with the Association stronger and more active than ten years previously, in a positive mood and a progressive state.

CHAPTER 9:
The Millennium and Beyond

An Opportunity for Education Funding

In 2001 Kenneth Edwards, a past Chairman of the Somerset BKA and a Senior Lecturer at Bridgwater College, drew Council's attention to a Government funded scheme from which the Association might receive considerable financial benefit. At a Council meeting held at Othery on 12th May of that year, Ken explained the scheme. He was confident that funding could be obtained in respect of beekeeping meetings organised by the Association and its divisions when training took place. The funding, controlled by the Further Education Funding Council (FEFC), necessitated our forming a partnership with Cannington College.

The SBKA Council decided that it should take advantage of such a scheme and Ken Edwards agreed to become the Special Funding Officer. He, supported by senior officers of the Association, put considerable effort into negotiating a contract for this partnership. This task proved to be unbelievably difficult and protracted, not only because of changes in the rules conditions and procedures, but also because during this time, Cannington College was experiencing severe problems within its management and communication was very difficult.

Somerton Division's Beginners' practical Class at Tengore House, Huish Episcopi, 2001.   *Photo: T. J. Harris*

Mike Milton conducting an Introductory course at Somerton, Winter 2001. *Photo: G. Fisher*

Meanwhile, at many of the Association's meetings, members were signing up to learning agreements and attendance records were being kept. Caroline Butter, the Association's Education Officer was responsible for this time-consuming part of the procedure. Progress of the scheme was an agenda item at Council meetings for over two years with no certainty of the outcome. Flaws were then found in the initial contract and this had to be re-drawn. Council eventually agreed on how the funds, if they materialised, would be dispersed between the Association and its divisions. Because of the delay, which lost the Association benefit for a year, Cannington College agreed that a larger share of the income would be allocated for the academic year 2002 – 2003. The funding body changed from being the FEFC to the Learning Skills Council (LSC), with consequent changes of rules governing the scheme and causing further delay.

Eventually, all the effort and perseverance paid off. The submissions, learner agreements and enrolments were submitted for the year 2002–2003 and the total sum received by the SBKA was £24,893.55 (£10,770.85 allocated to SBKA and £14,122.70 to divisions).

Following a number of poor Ofsted inspection results Cannington College merged with Bridgwater College, becoming a large department within it, re-named as the Cannington Centre for Land-based Studies. There was a complete change of management and few were familiar with procedures and developments so far.

New procedures were put in place and once again all of the submissions were made this time for the year 2003–2004. The total income was £20,041.36 (£7,779.02 allocated to SBKA and £12,262.34 to divisions).

The first payment was received in two instalments in the spring of 2004 and a further payment was received later in the year in respect of 2003–2004. Total receipts to date have amounted to a two-year total of nearly £45,000, apportioned, as agreed in Council between the Association and its Divisions, with the proviso that support would be given to weak divisions from the central fund.

The future of this arrangement and source of income is uncertain, although the contract extends until 2007. However, the income is already having a great impact on what the Association is able to do in promoting the craft. The funding has not been obtained easily. Considerable administrative work by Ken Edwards, the Special Funding Officer, and the co-ordination of procedures by Caroline Butter, the Somerset BKA Education Officer must be acknowledged.

A Boost in Demand for local Honey

It is pleasing to be able to state that there has not been a failure in the honey harvest for some years now. For those who have cared for their bees and kept them free from varroa and other troubles colonies have generally produced a worthwhile harvest each year. 2003 was exceptionally good. A beekeeper at Compton Dundon took 238lb from one hive. One wondered when the flow of nectar would cease, and to crown this excellent summer flow, bees on Exmoor also did well on the ling.

Alex Morrice and David Charles discuss the Exhibits at the Somerton Division Honey Show and Craft Fayre, Huish Episcopi School, 2003. *Photo: T. J. Harris*

There was no problem of having a glut of honey that could not be disposed of. It is an ill wind that blows no good, for a matter beyond our control boosted sales considerably and enabled a better price than that previously obtained. In 2003 the Food Standards Authority ordered all blends containing honey from China to be withdrawn from sale and supermarket shelves were promptly cleared. Minute amounts of an antibiotic known as chloramphenical had been found in test samples. The presence of this antibiotic in honey could pose a human health risk. Once weaned on to local honey, many consumers have stayed with it. In 2005 the average retail price of 454g local honey by the jar was £3.00.

The Beekeeping Scene

Since the registration of Bayvarol and Apistan, the control of varroa mites has been relatively easy. From the outset beekeepers were warned that sooner or later resistance would become apparent, especially if there was widespread misuse of the chemicals used. Alternative, but more complicated methods were encouraged, such as the use of formic acid.

Resistance to pyrethroid based acaricides through misuse was first observed in Italy in 1995, and in Southern France and Switzerland shortly afterwards. In 2003 resistant mites were found in North Devon, and misuse of chemical strips was blatantly evident in one case. In November of that year Richard Ball, the Regional Bee Inspector informed Council that colonies with resistant mites, belonging to a Dorset beekeeper, had been identified at Milborne Port and they were in a state of collapse. Occurrences in neighbouring counties prompted Council to take urgent action in alerting and educating the beekeepers of Somerset. In the spring of 2004 the Association in co-operation with Richard Ball and other R.B.I.s, ran four one-day workshops open to all beekeepers. These were well run, well attended and highly successful events.

Most beekeepers seem to cope with the changing situation regarding varroa mites. Sadly there has been an increase in incidence of European Foul Brood accompanied by concern that it may be removed from the list of statutory notifiable diseases. The increasingly mobile age in which we live brings threats of other, previously unheard of hazards to beekeeping.

Gerald Fisher promoting SBKA at the launch of Somerset Gateway (Somerset Food Links), 2004.

Photo: T. J. Harris

The Millennium and Beyond　　　　　　　　　　　　　　　　　　　　　　　　　　　95

Jake Wallis performs at Taunton Show, 2005.　　　　　　　　　　　Photo: K. A. T. Edwards

Beekeepers have already been alerted to the threat of the small hive beetle and two species of another parasitic mite known as *Tropilaelaps*. Under these conditions it is apparent that beekeepers need, and are at present, receiving support from the Department for Environment, Food and Rural Affairs, (DEFRA) through the literature published and through the Inspectorate, currently under threat of being reduced.

Regarding the DEFRA proposed cuts to the Bee Health Programme, a small delegation of Taunton members met with their Member of Parliament, Jeremy Browne MP in early August 2005 to discuss the proposed cuts with him and received a sympathetic ear. His interest was such that he accepted an invitation to visit the Heatherton Park apiary at the next meeting. He was duly kitted out

Stella Fisher promotes the products on the SBKA stand, National Amateur Gardening Show, 2005.

Photo: Richard Hudd, Western Daily Press

for beekeeping, and was impressed with his visit, which created valuable press publicity. Later in the year, at the instigation of Caroline Butter, representatives of four Divisions having members within the Wells Constituency, met David Heathcote-Amory MP regarding the same matter. Subsequently, in his diary published in the Central Somerset series of newspapers, after referring to his choice for party leadership, expressing his views on ID cards and other matters, Mr Heathcote-Amory stressed the value of bees. He referred to the proposed cuts and stated that he would write in strong terms to DEFRA on behalf of the local Beekeepers' Association. After consulting the Charity Commission, Somerset BKA took a strong lead by donating the sum of £500 from the Cannington Funding income to the BBKA's diminished fund for use in fighting the proposed cuts.

Taunton and District Division's "Rent-a-Hive" Scheme

This scheme, operating from the Heatherton Park apiary, commenced in 2005 following a six-week beginners' course held during the early spring. Prospective beekeepers joining the scheme are obliged to sign a contract with the Division undertaking membership for a minimum of two years. In return they receive a five-comb nucleus which is kept at the apiary. They tend these as if their own, but under the supervision of experienced beekeepers.

Left to right: Andy Hutton, Gay Brown and Richard Capstick tending their "rented hives" under supervision at Heatherton Park, 16th July 2005. *Photo: A. D. Charles*

The Millennium and Beyond

At the end of the year, with a season's guidance and experience behind them, they have the option of transferring the colony to their own hive and taking them to their own apiary, paying for the frames and foundation used. If they feel beekeeping is not for them, the stock remains at Heatherton Park, being the property of the Division.

Membership: an upward Trend

It is encouraging that since the year 2000, with the exception of 2003, there has been a growth in membership. In 2000 there were 367 members, including honorary members. In 2005 there are 434, having jumped from 413 in 2004. This must be attributed to the hard work undertaken by officers of the Association both centrally and at divisional level. The various programmes of lectures, demonstrations, introductory and improvers' courses and special events illustrate that the Association, although smaller than in former times is an active and growing one and prospects bode well for the future. Long may the honeybees produce delicious Somerset honey and continue to pollinate the apple orchards, agricultural crops, gardens and wild flowers of Somerset.

Analysis of Membership

Division	1980	1981	1982	1983	1984	1985	1986	1987	1988	1989
Bridgwater	52	53	57	52	43	37	35	30	28	18
Burnham	27	21	28	16	14	15	14	15	19	20
Central	28	28	27	18	17	17	16	17	15	12
Exmoor	76	68	64	62	59	59	54	47	49	48
Frome	16	23	20	19	14	11	10	7	10	9
Mendip	101	101	101	89	96	93	64	68	74	54
Somerton	73	76	77	79	71	72	66	56	45	66
South Eastern	40	42	41	43	39	39	37	35	30	31
South Western	97	76	55	59	50	33	40	35	36	42
Street & Glastonbury	74	67	62	37	38	26	30	25	28	24
Taunton	108	99	99	96	88	90	81	67	56	68
Wedmore				22	24	24	23	19	19	19
Yeovil	28	25	20	15	21	20	14	18	20	27
Total	**720**	**679**	**651**	**607**	**574**	**536**	**484**	**439**	**429**	**438**
Average	*60*	*57*	*54*	*47*	*44*	*41*	*37*	*34*	*33*	*34*
Hon Members (with bees)									16	13
Hon Members (without bees)										
Grand Total	**720**	**679**	**651**	**607**	**574**	**536**	**484**	**439**	**445**	**451**

Division	1990	1991	1992	1993	1994	1995	1996	1997	1998	1999
Bridgwater	31	33	30	26	22	19	18	11	0	0
Burnham	25	19	14	16	13	20	18	16	14	14
Central	14	15	17	15	10	13	10	8	7	9
Exmoor	51	47	41	40	46	42	39	36	33	33
Frome	6	11	9	12	14	13	12	13	12	13
Mendip	67	60	53	43	46	43	46	44	40	41
Somerton	62	57	56	60	49	53	61	59	78	85
South Eastern	35	41	37	37	33	30	30	33	36	27
South Western	39	35	40	32	27	26	24	20	16	15
Street & Glastonbury	28	27	20	15	12	13	8	7	0	0
Taunton	69	67	78	82	70	72	64	68	65	60
Wedmore	18	23	24	21	24	21	19	19	15	11
Yeovil	22	28	33	28	23	25	23	23	24	22
Total	**467**	**463**	**452**	**427**	**389**	**390**	**372**	**357**	**340**	**330**
Average	*36*	*36*	*35*	*33*	*30*	*30*	*29*	*27*	*31*	*30*
Hon Members (with bees)	13	12	9	8	11	7	7	7	12	9
Hon Members (without bees)									7	7
Grand Total	**480**	**475**	**461**	**435**	**400**	**397**	**379**	**364**	**352**	**339**

Division	2000	2001	2002	2003	2004	2005
Bridgwater	0	0	0	0	0	0
Burnham	12	9	9	10	12	13
Central	8	7	7	7	8	12
Exmoor	31	30	30	28	25	29
Frome	13	12	18	19	16	25
Mendip	41	36	38	35	38	29
Somerton	98	107	109	106	120	113
South Eastern	22	23	20	25	28	25
South Western	13	10	11	12	21	24
Street & Glastonbury	0	0	0	0	0	0
Taunton	71	73	82	80	82	93
Wedmore	12	21	23	20	21	26
Yeovil	24	25	30	27	30	33
Total	**345**	**353**	**377**	**369**	**401**	**422**
Average	*31*	*32*	*34*	*34*	*36*	*38*
Hon Members (with bees)	17	19	10	10	11	11
Hon Members (without bees)	5	4	7	6	1	1
Grand Total	**362**	**372**	**387**	**379**	**412**	**433**

Analysis compile4d by Mrs Sharon Blake, Treasurer

CHAPTER 10: Presidential Profiles

From its inception in 1906 until the present time, the Association has elected a total of fourteen presidents, (see page 152). For most of the Association's existence presidents, once elected, were re-elected year on year, generally remaining in office until their death. The exception was Miss Bindley. Many prominent members in the past were deserving of this office, but a vacancy rarely arose. Who would have had the temerity to nominate someone to stand against the great T. W. Cowan or L. E. Snelgrove? Indeed, who would dare to offer themselves? It would have been regarded as very bad form indeed!

In August 1983 a new set of rules was adopted and rule 10 has from that time stated *"anyone who has been elected President or Chairman for three consecutive years shall not be eligible for election as such for the next succeeding year."* This rule has enabled members to elect several presidents over the last twenty-two years. The author has spent considerable time in researching their full, interesting and diverse lives and work and hopes that the reader will find them interesting.

Lady Smyth, first President, 1906–1914

Formerly Mrs Emily Frances Edwards, a society beauty, she was the cousin of Sir Greville Smyth, Bart with whom she had had a long affair. They eventually married in 1884, he adopting their daughter Esmé whom she had borne him some years earlier. Their home was Ashton Court, Bristol, the Smyth family seat for four hundred years. For much of each year they were travelling abroad or were in residence at their country seats. These included Ardross Castle in Scotland, the Italian lakes and Nuwera Eliya hill station in Ceylon. It was Claridge's for the London season. Lady Smyth was on friendly terms with the Prince of Wales, later to be King Edward VII who was known to have made a number of visits to Ashton Court over a period of many years.

Lady Emily Smyth
Photo courtesy of the Malago Society, Bristol

Sir Greville died in 1901. On becoming a dowager Emily liked to be known as Dame

Emily Smyth. There is no evidence that she kept bees or had them on the estate, though the latter is probable. However, she did have a great friend, Miss Helen Dawe, lady superintendent of the Hill House Home for Gentlewomen, founded and funded by Lady Smyth. Miss Dawe was Secretary of the Bristol, Somersetshire and South Gloucestershire Bee-keepers' Association for five years from 1895. In 1898 Sir Greville's sister, Florence wrote in her diary 1st October *"After luncheon drove with Miss Dawe, who really is a most interesting person on bees and poor people".* It may well have been Miss Dawe who prevailed upon Lady Smyth to accept the Presidency of the Bristol, Somersetshire and South Gloucestershire Association, a position she held for the seventeen years of its existence prior to being elected president of the newly formed Somerset Association in 1906.

In those days institutions desired the president to be a person of eminence, wealth and influence as a figurehead. There is little evidence of Lady Smyth's attendance at the Association's functions or of active involvement in any way. She was frequently away, but not so much as before the death of her husband. However, she suffered greatly with arthritis which curtailed her activities during her final years. During this Edwardian period Lady Smyth laid out considerable sums on material and lasting benefits for the people of Long Ashton in particular and of Bristol in general. For example, she built the Long Ashton Almshouses in 1903 at a cost of £2,500. Despite their impecunious state there is no evidence that this Association or its forerunner benefited much from her generosity. As President she subscribed two guineas (£2. 2s. 0d or £2.10 in new money) compared with five shillings (25p) for gentry and half a crown (12.5p) for cottagers. She also donated a pound each year as first prize in the premier class of the Association's show of bees and honey. This was for the finest collection of honey and wax staged in the most attractive form on a space not to exceed three feet by three feet.

Lady Smyth was President of the Somerset Beekeepers' Association from its inauguration in 1906 until her death, aged 79, in November 1914. The history and demise of the Smyth family and the Ashton Court estate is really fascinating, but it would be inappropriate to expound further here. Those interested are referred to the Malago Society and the books on "The Last Smyths of Ashton Court" by Anton Bantock, ISBN 0 9507813 6 3.

Thomas William Cowan, F.L.S., F.G.S., F.R.M.S., F.E.S., D.Sc.,Ph.D., second President, 1915–1926

In complete contrast to his predecessor, T. W. Cowan was a highly skilled and knowledgeable beekeeper. He was a skilled engineer, geologist, archaeologist, administrator, writer, lecturer, microscopist and a remarkably able linguist. Coupled with boundless energy, enthusiasm and longevity this versatile, talented, yet modest and unassuming man made a tremendous impact on the beekeeping scene over many decades. Unlike his successor,

Snelgrove, by the time he came to Somerset his major work was already accomplished.

Cowan was born at St. Petersburg in 1840, the son of a Scottish civil engineer in the service of the Russian Government and with the rank of a general. He was educated in Russia and on coming to England he trained as an engineer at the London School of Mines. His family owned the Kent Iron Works at Greenwich.

As a young man he built one of the first steam carriages used upon English roads with which he achieved a speed of 20 miles per hour, but the passing of the Locomotive Act, with its many restrictions and the limit of speed to four miles an hour put an end to his pursuit in that direction. In later life he became skilled at wood carving.

T. W. Cowan

At the age of only 22 Cowan had taken over the business from his father. For much of his adult life he lived at Horsham, Surrey, where he had built a large house with extensive grounds planted with bees in mind.

Cowan's life was largely devoted to beekeeping. In his early beekeeping days he was a follower of Thomas White Woodbury, Devon, and initially used Woodbury hives and the Ligurian bees favoured by Woodbury, which he much preferred to the English black bees used by the skeppists. He later designed his own hive, "the Cowan hive". He was deeply involved in association work at all levels and became internationally recognised. A man of great renown, he was held in the highest regard throughout the craft. In short, he was a giant in beekeeping; one of the greatest beekeepers of his time whose influence was pivotal in the way that bees are managed today. For instance, he was Chairman of the British Bee-keepers' Association committee that in 1882 adopted the British Standard frame, which, incidentally is almost identical to the Woodbury frame.

Author of several books covering all aspects of the science, the craft and the products, his best known work is "The British Bee-Keeper's Guide Book" first published in 1881 while he was living at Horsham. Prior to this there was not in England any reliable work on beekeeping at a price within the reach of ordinary people interested in the craft. It was translated into no fewer than seven languages including Danish, French, German, Russian, Spanish and Swedish. The greatly revised 25th edition of 1924 was still in print at the time of his death. "The Honey Bee – Natural History, Anatomy and Physiology" published in 1890 was the standard reference for many years. "Wax Craft", published in 1908 was the

Portrait of T. W. Cowan from the collection of the Somerset Archaeological and Natural History Society

only comprehensive book on this subject for over seventy years and was translated into Russian and French. For over forty years Cowan was proprietor and editor of the now defunct British Bee Journal that in 1885 he had taken over on the death of the Rev. H. R. Peel and made it a weekly publication. He subsidised the Journal from his own pocket for the benefit of beekeepers until it was in profit. He published, in instalments, a history of the BBKA representing fifty years of bee-keeping progress. After his death this was reprinted as a book by BBKA to commemorate its jubilee, (fifty years). Until 1943 the British Bee Journal was the official organ of the British Bee-keepers' Association.

It was largely through T. W. Cowan's efforts that the British Bee-keepers' Association was founded and he worked for it from its inception, being elected to the committee at the first annual general meeting in 1874. Soon he was elected Chairman and amazingly held this post continuously for forty-eight years.

In 1886 a huge Indian and Colonial Exhibition of honey was held in the vast conservatory adjoining the Albert Hall at South Kensington and a two day conference of beekeepers was held in conjunction with this. Cowan presided and also delivered a lecture entitled "The Development of Bee-keeping as an Industry". Resulting from all this, Cowan was invited, as a representative of British beekeepers, to visit Canada, which he did the following year, visiting America at the same time. These were indeed golden and high profile years.

In 1898 the family emigrated to California where he established a son in fruit farming. This move brought tragedy to the Cowan family, for a son and daughter, travelling to join them aboard the steamship *Mohegan* were drowned on 13th October 1898 when it ran on to rocks at the Manacles off Cornwall. The Cowans bore this loss *"with exemplary Christian fortitude."*

In his early sixties, probably in 1903, he came back to England and settled in Somerset. He lived at Bishop's Hull, Taunton, the propitious timing of his arrival enabling him to become a founder member and a Vice-President of the newly formed Somerset BKA. He was President of the Taunton Division and despite his national and other commitments he was an active member, judging, lecturing and advising beekeepers. He was also President of the Taunton Field Club.

Later in life Cowan was unable, for various reasons to attend BBKA meetings regularly and it has been said that it would have been better for the Association had he not continued as Chairman for so long. New blood may well have been beneficial but he was certainly not "hogging the post". He was absent from the BBKA annual general meeting in March 1915 owing to the serious illness of Mrs Cowan, but in the previous month he had written a letter to Mr Reid, the Vice-Chairman saying *"…I have on several occasions asked the Council to accept my resignation, but have conceded to their kind wish for me to retain the position a little longer. Now that I have served the Association in this capacity for more than forty years, and*

The Cowan family in their home at Clevedon, circa 1925

as I am over 75 years of age, I feel that this is the right moment to retire, and I would therefore ask the Council to kindly accept my resignation…" Mr Reid proposed that the resignation be not accepted and the resolution was carried with loud and continued applause. So, he continued in this office until in 1922, when at the age of 82 he became BBKA President. After the death of Baroness Burdett-Coutts in 1906 this office was held only by Masters of the Worshipful Company of Wax Chandlers *ex officio*, but they were raised in status to Patron to make way for Cowan to be honoured with the Presidency. This arrangement continues today.

After World War I the Cowan family removed to Clevedon where he spent his remaining years still actively involved in the craft, his association work at national and local level and with the British Bee Journal of which he was still the proprietor. At Clevedon he was acquainted with Miss Freda Strudwick, M.A. who also lived there. She had been a member since World War I and was shown in the 1925 Year Book as being one of only three BBKA experts (final exam.) within the Association, the other two being Messrs Bigg-Wither and Snelgrove. Miss Strudwick later moved to Carhampton and her books on beekeeping at some time went to the Somerset Farm Institute (later Cannington College) library. In 1991, while David Charles was based there as County Beekeeping Adviser, he retrieved several from a tea chest of discarded books. These included a 25th edition of the Bee-keeper's Guide Book inscribed *"Miss Strudwick with Thos Wm Cowan's kind regards"*. Pasted inside the front cover is a previously unpublished family photograph showing T. W. Cowan with some books, Miss Cowan (left) and Mrs Cowan, (centre) at their home, Sutherland House, Clevedon.

Sutherland House, Clevedon — Photo: A.D. Charles

Cowan possessed a vast library of over 1,800 volumes consisting solely of literature devoted to bees and allied interests. Sadly, it was while reaching for a book in this library at Clevedon that he fell from his stepladder and died of internal injuries six days later on Whit Sunday, 23rd May 1926 aged 86. So ended the long and fruitful life of this remarkable man, the like of whom we shall never see again.

The funeral, at the Church of St. Andrew, Clevedon, on 26th May, was an occasion befitting the man. The SBKA was represented by Messrs Beck, Bradbury, Snelgrove and William West. He, and later his wife and daughter, were interred there, to the north-east of the church overlooking the Severn Estuary. The grave is marked with a substantial Celtic cross of granite, which over the years has inclined. The epitaph reads: *"Death is swallowed up in Victory"*.

L. E. Snelgrove

Louis Edward Snelgrove, M.A., M.Sc., F.R.E.S., third President, 1927–1965

Born in 1879, Mr Snelgrove originated from Sutton Veny, near Warminster where his father was a blacksmith. He was educated at Culham College, Oxford and Bristol University. Mr Snelgrove became an assistant master at Weston-super-Mare Central School for Boys in 1900. In 1909 he became headmaster and in 1914 was appointed Inspector of Schools for Somerset Education Committee, a position he held for 31 years.

After several years as an evening student at Bristol University Mr Snelgrove went to Queen's University, Belfast, as a graduate. In 1907 he took an honours science degree in physics and chemistry and in 1920 was awarded a M.Sc. degree for research on the etiology of the then named "Isle of Wight Disease of Bees". As a result of his research in Latin literature relating to Roman Bee Lore he obtained his MA degree in 1922.

Familiar with beekeeping since boyhood through helping his father manage eighty stocks Mr Snelgrove started beekeeping on his own account in 1902. He joined the Bristol, Somersetshire and South Gloucestershire Association and in July 1905 he became Honorary Secretary and Treasurer of this Association, which in the words of the committee was *"in a very backward state"*. As you will read in greater detail elsewhere in this book, Mr Snelgrove worked unceasingly with great drive and foresight to restore its fortunes. He was the kingpin and driving force in reconstructing it as the Somerset BKA, which he greatly fostered for the rest of his life and moulded into the Association whose centenary is now commemorated.

Mr Snelgrove resigned the secretaryship in 1912 due to pressure of his professional duties but he remained on the SBKA Council. A major project he undertook in the intervening years before he became Chairman, was a census of beekeepers, hives and colonies throughout the 550 parishes in Somerset in 1918-1919. He also did most of the work in drawing up a re-stocking scheme after the majority of colonies had been wiped out by the "Isle of Wight disease". In 1921 he succeeded Colonel Jolly as Chairman, a post he held until 1927 when he succeeded T. W. Cowan as President. His presidency continued until his death in 1965. He had held this office for nearly forty years, but in all gave sixty years of dedicated service to this Association and to the craft showing fine qualities of leadership and example.

Of his beekeeping, one immediately thinks of the Snelgrove board, his "water" method of queen introduction, the "one-hour" method of queen introduction, the BHS method of queen rearing and a multiple queen cage, the idea of which he was the first to demonstrate in public. He certainly brought a fund of practical ideas to the craft and a great deal of sound research, much of which has been left to posterity through his well known literary works Of his books "Queen Rearing"(1946) is regarded as the most complete work on the subject. Together with "Swarm Control and Prevention"(1934) and "Queen Introduction"(1940) this is a trio that must be on many a beekeeper's bookshelves.

In 1954, in recognition of his immense service to beekeeping the BBKA were pleased to honour him with honorary Life Membership and in 1956 he was elected its President.

"It is with profound regret and a deep sense of irreparable loss that the Council has to record the passing of our beloved President on 21st November, 1965." Leslie Hender, Chairman, SBKA.

"We always welcomed his visits because he listened to our difficulties with the greatest of patience and understanding." A Somerset headmistress.

"His humbleness was his greatness." Cecil C. Tonsley, Editor, British Bee Journal.

Leslie Hender (right) with Mr Hebden, CBI for Yorkshire, at Heatherton Park, 1961

Leslie Hender, N.D.B., F.R.E.S., fourth President, 1966–1970

A Devonian by birth, Leslie Hender joined the staff of Lloyds Bank in 1909 at South Molton and then Modbury. During World War I he joined the colours of the Royal Artillery and became an instructor in the School of Gunnery, his health debarring him from active service overseas.

On demobilisation he rejoined the bank at Modbury, later moving to Tavistock. This was followed by a short term in the Taunton office as accountant before being appointed sub-manager at Torquay. From thence he became manager of South Molton and in 1937 he returned as manager to Taunton, a position he held until his retirement in 1953.

During World War II he served with the Home Guard. He also became actively interested in every aspect of amateur dramatics. With the Thespians he played many parts in their wartime and immediate post-war productions and eventually became their chairman.

It is not known exactly when this man of many parts took up beekeeping. He became a member of the Taunton Branch and is first listed in the 1944 Year Book. His wife, Lilian became interested too and she joined a year later. They shared this interest for the rest of their lives, but each had their own stocks.

In retirement Leslie became energetically immersed in beekeeping, although long before retiring he was actively engaged as an executive member of the Somerset BKA. By the late 1940s he had passed the BBKA Senior examination and obtained the lecturer's certificate. In 1952 he became a BBKA honey judge, going on to attain the highest qualification of all, the National Diploma in Beekeeping. For some years he represented the Association on the former BBKA Council and was a member of the Examinations Board.

Leslie was a great honey exhibitor and was regarded as one of the country's foremost judges. Always a loved and popular figure at shows throughout the country, under his chairmanship the honey section at the Bath and West Show reached a very high standard. For some years he was on the executive of the National Honey Show and was also a member of the Examinations Board of the BBKA.

In the 1950s he played a key role in the establishment of the Taunton Division's apiary at Heatherton Park.

As a great supporter of the Young Farmers' Club movement, Leslie, from 1937, kept close contact with the Taunton Club, serving on the Advisory Committee as Chairman, and was later honoured with its presidency. He was keenly interested in village life. At one time he served as representative on the Rural District Council and was Vice-Chairman of the Finance Committee. In his later years he became an enthusiastic wine-maker and was one of the founder members of the National Guild of Judges. The Somerset County Cricket Club honoured him with life membership in recognition of his services as their Treasurer. He was a great sportsman and a countryman at heart.

He was Chairman of SBKA for many years prior to the death of Mr Snelgrove, after which he was elected President, a post he held for nearly five years until dying peacefully at his home in Lydeard St. Lawrence on 21st November 1970, aged 78.

Miss Mildred Duncombe Bindley, fifth President, 1971–1981

Bells rang out in 1889 for the birth of Mildred Bindley at Ixworth, Suffolk, where her father was curate. Her first birthday was celebrated in mid-Atlantic on the way to Barbados. Until the age of thirteen Mildred and her brother were cared for by a German nurse. They spoke German between themselves, hence her sound knowledge of the language.

In 1910, when Mildred was twenty, her father retired from Barbados and took a living at Hedenham, Norfolk. The lady next door was a beekeeper and Mildred became interested. Her father became Rector and Archdeacon in Norwich Diocese, which necessitated moving to Denton near Harleston. During World War I Mildred joined the Red Cross and trained as a nurse but gave it up after catching diphtheria, chickenpox and measles.

In 1918 she enrolled for a three-year course in horticulture at Swanley Horticultural College, Kent, taking beekeeping as an extra subject. W. Herrod-Hempsall was the examiner.

After two years at college and a third of practical experience she ran the large rectory garden commercially. She was awarded the Swanley College Certificate and the Royal Horticultural Society's first class certificate. By now she had her own hives and kept bees for the next sixty years.

In 1921 Mildred commenced work in England and France with Brenda Colvin, a garden designer. After the death of her father in 1931 she moved to Gloucestershire, then to Hinton St. George, Somerset where she was associated with John Barnwell in running a silver fox farm. These were good years, but when World War II came in 1939 John was called up at once. There were difficulties in running the farm single-handed and obtaining food for the foxes was problematic so the farm was sold up.

Mildred Bindley

During the war period Mildred promoted the "Dig for Victory" campaign growing lots of fruit and vegetables, encouraged people to keep bees, and she joined the ARP.

In 1946, Mildred moved to Dry Close, a lovely old house at Misterton near Crewkerne. She worked with Dr Eva Crane to establish the Bee Research Association in 1949, which later became the International Bee Research Association. She helped to catalogue the publications that were the foundation of the present library, and also made translations from German, and produced abstracts for "Bee World". She toured Germany with Dr Eva Crane in 1951, visiting numerous beekeeping establishments including the Beekeeping Institutes at Celle near Hanover and Erlangen near Nurnberg. A highlight was visiting Professor Karl von Frisch at the Zoological Institute of Munich University. Here she saw films of his experiments on the bees' senses of smell, colour and taste, and his latest film on the dances of bees.

A great traveller, Mildred attended Apimondia Congresses at Moscow and Grenoble. Apart from tending her bees both at Misterton and in the Cotswolds she was immersed fully in village life. She was a member of the Parochial Church Council and the Womens' Institute, and was on the Board of Managers of the local school.

Miss Bindley joined Somerset Beekeepers' Association in 1934 and almost immediately became Branch Secretary. She passed the advanced examinations of BBKA and was Secretary, Treasurer and complete organiser of the South-Western Division for about forty years. She was a member of SBKA Council and Chairman for three years in the mid-fifties.

She lectured and demonstrated widely, and as Education Secretary, organised numerous conferences and weekend schools at Dillington House and Cannington. She arranged for candidates to take their beekeeping examinations and conducted or invigilated numerous examinations herself. Another task Mildred undertook for many years until 1973 was the editorship of the SBKA Year Book.

Nominated to continue as President, Mildred surprised the Annual General Meeting held at Castle Cary in 1981 by withdrawing her nomination due to failing health and eyesight that meant she was no longer able to drive her car and play a fully active part. She had moved to a smaller house in Misterton and eventually to a nursing home. After Jack Lewis moved from the area she succeeded him as President of her beloved South-Western Division for the ultimate year of her life. She passed away peacefully on 2nd December 1983 aged 94. Her knowledge and long experience were immense, as were her contributions to SBKA in so many ways over nearly fifty years.

Rex Walter Sawyer, B.Sc., Hon. Fellow, BBKA, sixth President, 1981–1983

Born and brought up in East Anglia, Rex graduated from London University with a B.Sc. degree and embarked on a teaching career, first at a preparatory school in Minehead, then at a public school in Bombay. On returning to England he taught briefly at Sexey's School, Blackford, near Wedmore before taking up a post at Ilminster Boys' Grammar School. During World War II his interest in radio and flying resulted in a commission in the Royal Auxiliary Air Force as signals officer concerned with the development of night fighter airborne radar interception.

On demobilisation Rex returned to Ilminster where he became Head of Science. A keen all round sportsman, Rex was a great cricketer and coached the school team which was almost invincible. He was also a good rifle shot, a radio ham and founder of the School Radio Club. Most importantly for our craft, he took up keeping bees, developed a commercial apiary and an interest in melissopalynology. Whatever his knowledge, skills or interests he was ever willing to pass these on to those wishing to learn.

By 1950 Rex had passed all the BBKA examinations to Senior level. He successfully ran a large number of colonies in the Ilminster area and soon became actively involved with the work of SBKA. By the mid-fifties he was vice-chairman, then chairman in 1965 following the death of Mr Snelgrove. This post he held until 1980, leading the Association through some very difficult times. The breaking away of four divisions to form the Avon County Beekeepers' Association and the loss of the services of a part-time County Beekeeping Officer were just two problems during that time. He was the first President under the new rules that restricted him to a three-year term of office. In fact, he declined nomination for a third year. Subsequently, until his death he was a Vice- President and Honorary Member,

Rex Sawyer in 1989 after his second book was published. Rex is holding the silver medal awarded to him at Apimondia, Rio de Janeiro, for his book "Honey Identification".   *Photo courtesy of Mrs. Enid Sawyer*

continuing to play an active part in our affairs. Rex was also instrumental in the formation of the South West Counties Joint Consultative Council, an influential forum that serves a valuable function to this day.

As a botanist and a beekeeper it was not surprising that Rex should become fascinated in a scientific aspect of our craft. His training, ability and aptitude equipped him to become expert in the identification of pollen grains in honey, a specialist science known as melissopalynology that has a practical application in the determination of sources of nectar by the incidental grains present in it and subsequently in the honey.

In the early 1970s there was widespread concern among beekeepers about cheap foreign honey being labelled and sold as English. His pursuit of the vendors of foreign honey labelled as English did much to protect beekeepers from unfair competition and consumers from being cheated.

The BBKA asked Rex if he would run a weekend course for suitable beekeepers to be trained in pollen identification and the first, at Sparsholt College in 1976, was oversubscribed. In all, six were held in various parts of the country. In 1986, in recognition of thirty years of work on this specialist subject, Rex was elected an Honorary Fellow of the BBKA. During this time he had built up an extensive bank of pollen slides for reference purposes, carried out numerous examinations for beekeepers, public analysts, local authorities, honey packers and importers. He also examined pollen and honey for the development and control of a honey production unit in the Yemen Arab Republic.

Rex wrote many articles on melissopalynology for beekeeping journals and for scientific bodies, for example – the Association of Public Analysts. His first book, "Pollen Identification for Beekeepers" received excellent reviews in leading scientific journals and was awarded a bronze medal at the Apimondia Congress at Budapest in 1983. In 1988 his second book, "Honey Identification" was published, and this was awarded a silver medal at the Apimondia Congress in Rio de Janeiro in 1989. Both are standard works on the subject.

From 1979 Rex was an Honorary Research Associate of University College, Cardiff and taught post-graduate students in the Bee Research Unit there. Several students whom he trained gained advanced degrees, their theses having been on pollen in honey. Kim Thomas, M.Sc. wrote hers on "Pollen in British Honeys", Joyce Amoako on "Honeys of Ghana" and Alex Banda on "Pollen of Malawi" to name but three.

Rex travelled widely in this country and abroad. He attended a number of Apimondia Congresses, one of which was in Tokyo, and on his way there he lectured in China. Returning via New Zealand he did a lecture tour at the invitation of the NZ Honey Producers and Packers Association who sponsored his visit.

Through his skill, knowledge and ability to teach, Rex added a new dimension to this unique science carrying it forward to a higher level. Through his influence not only has knowledge of melissopalynolgy greatly increased but also its application for the benefit of honey marketing. Through his teaching, his books and papers he has ensured that this will continue, not only in the United Kingdom but throughout the globe. He gained recognition as a world authority on the subject and when he retired from this work he donated his vast library of pollen slides to Cardiff University.

When Rex died peacefully in his sleep on 17th March 1999 aged 96 he had been a member of our Association for over fifty years.

Leonard Henry Septimus Gulliford, seventh President, 1983–1986

Len's presidency reflects an interesting change in the social pattern and breakdown of class barriers in society. Prior to his election presidents had been of the aristocracy, gentry, academics or white-collar professionals. Len, an ordinary working man, but a skilled tradesman, was the choice of members that one must attribute to his skill as a beekeeper, his work for the Association and the craft, and his great popularity with all who knew him.

Len was born at Clapton, a hamlet near Midsomer Norton in 1911 and here he lived for his entire life, moving only to a new house next door to his childhood home on marriage, the site being a wedding gift from his parents. He was a knowledgeable countryman.

Len Gulliford with a feral colony of bees.   *Photo courtesy of Mrs Ann Mogg*

When he was eleven years old Len's father, Bert, purchased two stocks of bees and together with Len learnt how to manage them. It proved difficult for they had no experience of beekeeping. There were many swarms but no honey in the early years. Furthermore the bees became very bad tempered and they nearly gave up when Len's mother was badly stung and for a couple of days sported two black and swollen eyes. Passers by were seen to flap their arms and remark that there must be a wasps' nest nearby, especially when the hives had recently been examined.

Bert belonged to the Eastern Division of the Association. The visiting expert was Mr E. G. Hawkins of Wallbridge, near Frome and later from Rodden. When working in this area he travelled by train to Radstock, taking his bicycle with him, and then cycled to the various members who required visits. It was Mr Hawkins who showed Bert and Len how to re-queen and get the hives in good order.

Eventually Len and Bert became skilled in the craft and never looked back. Len joined the Association also. They became keen exhibitors and the numerous cups and trophies they won were always inscribed B. and L. Gulliford. Through his sister Len met Mary, a Dulverton lass then working near Bristol, and they were married in 1936. Mary became interested in the bees, but proved to be allergic to stings, so although Len and Mary worked together, Mary busied herself more with the honey and later on, when they were agents for Taylor's of Welwyn, with the sales. Always, though, beekeeping was a family affair and after Bert's death, trophies won at honey shows were inscribed L. and M. Gulliford.

Len had served an apprenticeship as a carpenter and worked for the firm of Wines in Midsomer Norton. When World War II broke out Len failed his medical for the armed services due to psoriasis, but found himself deployed on war reparations work. He spent several years working in London, living in lodgings and coming home when he could. His father tended the bees during this time.

When family life was restored after the war Len built up his beekeeping and became very involved with the craft and with the Radstock and Midsomer Norton Branch. In the late fifties he was an advisory officer and in 1972 he became Hon. Secretary and Treasurer of the Branch. After many years of service in this capacity Len eventually became president of the group, now re-defined as Mendip Division. Len and Mary supported and helped run many of the honey shows in the county and beyond. Then, for 25 years they took on the task of organising the Bees and Honey section for the Royal Bath and West Show.

In the early 1970s Len advised for the BBC2 Television series "Living in the Past" concerning a project where two or three families lived for some time in a pseudo iron age village on Cranbourne Chase. The inhabitants kept three wicker hives of bees as part of their domestic economy. Len stocked one of these hives and advised on their management. In 1980 Len featured with Angela Rippon in the making of a programme about bees for the BBC 2 Television series "Out in the Country".

Len and Mary Gulliford with a prize-winning exhibit.   *Photo courtesy of Mrs Ann Mogg*

Len died in 1991 aged 80. At Haycombe Crematorium, on the outskirts of Bath, the number of mourners present to bid farewell to this genial countryman and beekeeper was so great that the chapel was packed to the doors and many mourners were obliged to stand.

Alfred David Charles, eighth President, 1986–1989 and tenth President 1992–1995

Born in 1935, David showed a keen interest in nature and gardening from an early age. He was educated at Altrincham Grammar School for Boys from where he embarked on a career in horticulture and worked for Manchester Corporation Parks Department for four years, broken by National Service in the RAF from 1953 to 1955. He trained as an electrical mechanic and was stationed at RAF Henlow, Bedfordshire. In 1958 he was accepted for a two-year course at the RHS School of Horticulture, Wisley, Surrey, where he gained a Diploma in Horticulture and the Chittenden Award for top student.

David then gave up horticulture in favour of a career in teaching, attending in 1960, a special one-year course at Newton Park College, Bath, where he trained to be a specialist teacher in Rural Science. While at college a latent interest in beekeeping further developed, and when in September 1961 he was appointed to a teaching post in Slough, he was pleased to find there was a school beekeeping club. David took over the running of this club within a couple of years.

David Charles

David commenced beekeeping in 1962 with three colonies built up from swarms. He joined Buckinghamshire Beekeepers Association, became a committee member and soon was Honorary Secretary of Slough Beekeepers' Society.

In 1967 he was seconded to Reading University to undertake a full time course leading to a Diploma in Education, followed in 1968 by the appointment to a Head of Department post at St. Dunstan's School, Glastonbury. It was at this point that David became a member of the Somerset Beekeepers' Association which has continued to our great benefit to the present day.

An activity of note for the thirty-five years from 1970, a welcome diversion from his profession and beekeeping activities, was membership of the Street based Strode Opera Club, where David sang tenor, performing mainly in the chorus and in minor parts.

In 1970 David became Honorary Secretary of the almost defunct Street & Glastonbury Division and by dint of hard work built it up into a thriving division over several years, the culmination being the splitting of the division to re-form the Wedmore and Cheddar

Branch. In 1971 David become a delegate to the SBKA Council and in 1973 took on the editorship of the SBKA year book.

In parallel with his efforts at branch and county level, David had been busying himself with study for the BBKA examinations and in 1972 passed the Senior Certificate examination and was also awarded the prestigious Wax Chandlers prize for the top marked national candidate. In the same year he also passed his Senior Practical examination to become a Master Beekeeper.

The year 1975 saw an expansion of David's involvement into national beekeeping activities with his election to the BBKA executive committee, where he remained a very active member until 1993. During this period he was keen on improving communication with the membership and assisted Harrison Ashforth with the BBKA news notes. He was then responsible for creating the current BBKA News, which he edited for a number of years. The culmination of his time on the BBKA executive committee was his election as President for his final two years in office. Subsequently in recognition of his very considerable contribution to the BBKA, he was elected an Honorary Member.

Further activity on behalf of SBKA was undertaken in 1979 when David took over the post of Education Secretary from the long serving Miss Mildred Bindley. This entailed organising a number of weekend schools. The examination record book, which he inherited and still continues to complete meticulously for each candidate undertaking assessments, is an historical record of great interest in its own right. As Examinations Secretary he has been responsible for the organisation of all examinations over this twenty-five year period together with associated invigilation.

In 1989 and by now Senior Master at St. Dunstan's School, Glastonbury, David elected to take early retirement from teaching, but his beekeeping activities continued unabated. 1990 saw his appointment as part-time Somerset County Beekeeping Adviser, based at Cannington College, a post he held for four and a half years. During this period he established an apiary, ran educational courses and an adult bee diseases advisory service.

2002 marked ten years as a committee member of the National Honey Show, with Schedule editorship having been David's responsibility for much of this time. He was a keen exhibitor for many years.

In 1996 David was elected Chairman of the now expanded Somerton and District Division, his old division of Street & Glastonbury having previously amalgamated with Somerton Division. For the years from 2000 to 2003 he served as President of Somerton and District Division.

David has lectured, demonstrated and judged widely in this country for over thirty years and is therefore a very well known national beekeeping figure. He has in the past run up to twenty-four hives, but latterly manages about ten. His other activities have included the study of continental beekeeping and he has attended three Apimondia Congresses to date. David retains his interest in horticulture and still regularly lectures to various gardening clubs and societies and manages his own garden. I am also very pleased to record that he is the first of our surviving Presidents and currently enjoys moderately good health.

Patrick Rich, Chairman, Mendip & District Division

Clifford H G Langford, ninth President, 1989–1992

Known to colleagues and beekeeping friends alike as Sam, Clifford Langford was born in 1914 at Broadwindsor, Dorset close to Lewesdon Hill, an area he loved all his life. While still a small boy the cottage where he lived with his parents and two elder sisters was burnt out leaving them homeless. The family moved to Knapp, North Curry where they lived in grandfather's house, he moving in with other relatives. Sam was educated at Huish Grammar School in Taunton. On leaving school he served an engineering apprenticeship with Petter's, later to become Westland Helicopters. He lodged in Yeovil, and during World War II he travelled to work each day by motorcycle working on that renowned fighter plane, the Spitfire. As he was in a reserved occupation he was part of the Home Guard, covering Somerton area. He married Mary Fouracres and they set up home at the mill, Barton St. David.

Clifford (Sam) Langford

In the early fifties Sam and his family moved to Charlton Adam. It was at this time that he became interested in bees. In the adjacent orchard there was soon a row of hives and later the family Jersey cow that Sam milked each morning before going to work.

Sam's work at this time was at Ilton Aerodrome, Ilchester, where he worked on helicopters. On Saturdays he went to Knapp to help his mother and sister Gladys, who also kept bees at one time and was a member. Sister Dorothy emigrated to Trinidad with her husband, an orthopaedic surgeon. They kept bees there and the honey was sometimes used in the healing of patients after surgery.

Sam joined the Somerton Division around 1953 and was always a participating member, serving variously on the committee and as advisory officer. The family moved to Curry Rivel. By the 1970s Sam was Chief Inspector at Westland Helicopters at Yeovil, but his health was deteriorating due to diabetes and as a consequence he was able to take early retirement. There was more time available then for both the bees and Association work. Sam was a keen exhibitor of honey and mead, winning many prizes at the various shows in the county. He passed on his beekeeping skills to many new and less experienced beekeepers.

In 1974 he was elected Chairman of the Division, an office he held for many years and then in 1985 he became President. The greatest honour was paid to him in 1989 when he was elected President of the Somerset Beekeepers' Association, an office he held for three years.

Grandfather's house at Knapp had been divided into two homes during the 1960s, and in the 1980s he and Mary moved back, next door to sister Gladys. Many beekeeping meetings were held here where Mary was known for her superb teas.

Sam was a highly principled and conscientious man, always courteous and expecting the courtesy to be returned. He worked by the book and soon spoke out when things were not done properly or according to the rules.

A true countryman and knowledgeable on many subjects, he was much respected by his work colleagues and fellow beekeepers. Sadly he was struck down by a severe stroke and died nearly a year later, in March 1999, aged 85. His funeral at North Curry was attended by a large number of friends and fellow beekeepers, some having travelled from Devon to be present. He is remembered by the C. H. G. Langford Cup, presented by him for the best sample of light honey at the Somerton and District Divisional Show.

John Newcombe

John Charles Newcombe, eleventh President, 1995–1998

Few people indeed could lay claim to membership from such an early age and for such a long period of time as John Newcombe. He started keeping bees in 1948 and became a member of the Association's Exmoor Division.

John's early years were spent in Taunton, but his family removed to Minehead when he was a boy of ten and most of his life has been lived there since that time. As was the case with most young men, John completed two years' National Service in the early 1950s, serving in the army, and was stationed for some of the time in Germany. His father tended the bees while John was away from home. For the remainder of his working life John followed a career as a telecommunications engineer. In retirement he and his wife Sheila live on the fringes of the town with an out-apiary a short distance from home.

John has served both his Division and the central body in many capacities. He was delegate to Council from 1960 onwards and was the Division's hon. secretary from 1981 until 1993. He was then Chairman for five years until 1998. Over the years John has rendered help and encouragement to scores of inexperienced beekeepers. In the 1960s and 1970s he was one of several Seasonal Bee Disease Officers appointed to inspect colonies under the Foul Brood Diseases of Bees Order. He also taught beekeeping at classes held by the local evening institute.

In 1990 John was elected Chairman of SBKA and served in this capacity for three years, after which he was elected President in 1998. At this time he also received the West Country Honey Farms silver rose bowl awarded annually to the member who is adjudged to have contributed most to beekeeping and the SBKA.

In 2001, at the end of his term of office, John was elected a Vice-President and also was awarded a certificate by the British Beekeepers' Association in recognition of fifty years of membership and service to the craft. He is currently President of the Exmoor Division.

Anthony Creswick Elcomb, twelfth President, 1998–2001

Anthony Creswick Elcomb, Tony to his friends, was born at Kasauli, India, in 1924 while his father was serving there in the British Army. At the age of five he came to England and the family went to live with his grandmother and great grandmother in Sheffield. He was educated at nearby Dronfield Grammar School. Tony loved country life. In his youth he cycled many miles through the Derbyshire countryside and in his school holidays worked on a farm. *"I should like to have a farm of my own one day"*, he declared. In 1941, at the age of 17 he joined the Royal Marines, doing his training at Lympstone in Devon, and serving as a Marine throughout the war.

After the war he transferred to his brother's regiment, the King's Own Yorkshire Light Infantry and was at Staff College before a series

Colonel A.C. Elcomb
Photo courtesy of Mrs N. Elcomb

of postings overseas. The next 25 years brought postings to Cyprus, Kenya, Malaya, Berlin, and finally two years in Washington, USA before his retirement in 1981.

Looking to the future, Tony, true to his youthful ambitions, had purchased Home Farm at Stoney Stoke, near Bruton in 1965 in readiness for his new life at the end of a successful military career. *"I would like a cow and a hive of bees"* he stated at that time. The cow came the day after his retirement and calved the following day, so milking was immediately part of the daily routine. During the ensuing years he built up a fine herd of Jersey cows of which he was justly proud.

The bees came a little later, though Tony had joined the Association's South Eastern Division as early as 1978. He was an active member and soon came to grips with the craft, running several hives. Then, in 1986, when a new divisional Hon. Secretary was required Tony took on this role. He remained as Hon. Secretary of the Division for fourteen years.

During much of this time he had attended Council meetings and his potential as a future Chairman soon became apparent. He was Chairman from 1993 until 1996. Prior to Council meetings working lunches were the order of the day for senior officers. This enabled business to be conducted in a quick, slick and efficient manner. Tony could not abide procrastination and delegates knew that meetings would not be long and drawn out. After all, a herd of cows awaited his attention at Home Farm.

In 1998 Tony was elected President, but alas, eighteen months into his presidency he became terminally ill. His nature was such that he doggedly pressed on to the best of his ability, supported by his wife Nora who often drove him to meetings until he became too ill to attend. Tony's life ended, aged 76, on 13th March 2001, just four days short of completing his three years as President, leaving both the Association and his Division in a stronger state.

Neil Brian Trood, thirteenth President, 2001–2004

Somerset born and bred, Neil was educated at Huish Grammar School, Taunton, the Somerset College of Art and the Somerset Technical College. The whole of his working life was spent in the newspaper industry except for his period of National Service served with the Royal Air Force. During this time he travelled widely in the Far East.

A quiet and unassuming man, nobody would guess that Neil was in former times a playing member of both the Taunton Operatic Society and the Taunton Thespians, taking on character roles. He was stage director for many productions. He also performed with the liberty Players, eventually running the club as its stage director.

It was through the Thespians that during his teens Neil met Leslie and Lilian Hender and it was they who interested him in beekeeping. Later on, when Neil had moved to a rural area he was able to take up the craft. In 1978, he and his son Timothy attended one of John

Newcombe's beekeeping courses in Minehead. Neil joined the Association's Taunton Division in 1979, and was elected to the committee. He accepted, on a temporary basis, the post of honey show secretary for both the County honey show and the Taunton honey show held in conjunction with the Taunton Flower Show each summer at Vivary Park and he still most ably does this important job. Neil was also Hon. Secretary of the Division from 1989 until 1993.

As a member of Council he was instrumental in the re-introduction of the year book in 1991 and for many years was responsible for selling advertising space. Over several years he raised hundreds of pounds to offset the cost of production. In 1996 Neil became the Association's Chairman, an office he held for

Neil Trood

three years. In 2001 he served a three-year term as President. He is a recipient of the West Country Honey Farms Rose Bowl Award and a Vice-President of the Association. Both Neil and his wife Jenny remain active in Association work.

Mary Barnes-Gorell, 14th President, 2004 to date

From her early days of beekeeping in 1990 and membership of the South-Eastern Division, Mary has been a constant worker for and supporter of Somerset BKA. Within her Division she has supported not only the beginners but many beekeepers in need of help. This applied particularly to past President Tony Elcomb during his final illness, and from whom she took over the secretaryship. She has been a delegate to Council, taking on the role of Covenants and then Gift Aid Secretary, very time-consuming work that has resulted in great financial benefit to the Association. Mary exhibits in honey shows within the county and

Mary Barnes-Gorell

encourages others to do so. On the publicity stands staged by the SBKA Mary is often to be found meeting members and the public, promoting the craft whenever the opportunity occurs. She is looking forward to playing a full part in the forthcoming centenary events.

Mary Barnes-Gorell at the Association's Publicity Stand, Royal Bath and West Show, June 2005.

Photo: A. D. Charles

Chapter 11: Some prominent Members

Charles Tite

Born in 1842, Charles Tite was a native of Taunton and his early years were spent in this town. He was a man who started life with no initial advantages, owing his success in life largely to himself. When still comparatively young he moved to Yeovil to become editor of the *Western Flying Post*. This newspaper later amalgamated with the *Western Gazette*.

Charles Tite was an early member of the British Bee-keepers' Association. After his visit to Grantham referred to in Chapter 1, Cowan states *"The Show at Grantham suggested to Mr Tite that knowledge of bee-keeping would be most widely diffused by the formation of County Associations. Mr Tite was the originator of these important societies and advocated their formation in a letter that appeared in the "British Bee Journal" of September 1875. It was headed "Proposed County Associations of Bee-keepers".* Tite followed up his suggestion with further letters in the journal over the next few years, giving many valuable suggestions as to the rules and aims, and also hints for County Associations. In the British Bee Journal for November 1876 he issued the following challenge: *"If a score of your readers will do likewise, I will guarantee to obtain six new subscribers to the B.B.K.A. before the next annual subscription becomes due in May, or, if I fail, I will pay £1.10s. to our Hon. Secretary."*

Portrait of Charles Tite from the collection of the Somerset Archaeological and Natural History Society

In 1886 the newspaper was sold for a large sum and Charles Tite, now a wealthy man, retired from professional life. After a period of travel and a short sojourn at Wellington, Somerset, Tite returned to live in Taunton at Stoneleigh, South Road, where the next forty years were spent in pursuit of active and useful work for the causes he loved. He became prominent as an archaeologist and did an immense amount for the Somerset Archaeological Society. His masterpiece was the Tite library which consists of a vast collection of Somerset books, paintings, engravings, etc. housed, catalogued and arranged in the Tite room at the County museum. He presented many artefacts to the museum, notably trade tokens and coins, and in fact was a great benefactor to Taunton.

Charles Tite was at one time a magistrate, member of the Borough Education Committee and a generous supporter of St. Mary's Church. As chairman of the trustees of the Taunton Town Charities he became the constant friend and adviser of the inmates of Gray's and other almshouses. He was a generous donor to the Free Library and was responsible for its extension in 1912. Moreover, possession by the town of land adjoining Vivary Park is due to Mr Tite.

A lifelong friend of T.W. Cowan, Tite may well have been influential in the Cowans retiring to Somerset. Little is known of his beekeeping, but he was, at one time, a frequent lecturer and honey judge. He was a Vice-President of SBKA having been a member of this body and its forerunners for 56 years until his death in 1933.

Lt.-Colonel H. F. Jolly, Chairman, 1906–1922

Herbert Francis Jolly was the elder son of Mr Frank Jolly, the senior partner in the firm of Jolly & Sons, of Bristol and Bath. He was born at Bath in 1862, was educated at Christ's College, Finchley, and spent some time in Germany to complete his education. After gaining business experience with one of the leading wholesale firms in London, he joined his father in the business in Bristol, being joined later by his brother.

As a boy he was always keen on natural history and microscopic work, which was useful to him when he became a beekeeper later in life. He began beekeeping at Henbury, near Bristol, in 1895, and became a Vice-President of the Bristol, Somersetshire and South Gloucestershire Beekeepers' Association. In 1902 he was elected a member of the Council. In 1906, when the Association was re-formed Col. Jolly was elected chairman of the Council, a position which he held until his death. He put his whole soul into the work, doing much to bring the Association to a flourishing condition, and spared neither time nor money in advancing its interests. Although a man of retiring disposition, naturally his enthusiasm brought him into prominence, and with members of the Association he was extremely popular. He kept only a few hives, but with these he carried out many experiments, and as a microscopist did research work in connection with bee diseases.

He joined the Volunteer Force, the 2nd Gloucester Royal Engineers, in 1887 as 2nd Lieutenant, and received the Long Service medal after twenty years' service in 1907, when he retired with the rank of Lieutenant-Colonel. When war broke out in 1914 he was beyond active service age, but assisted as a voluntary driver in the ambulance section of the Red Cross movement.

Although the business absorbed much of his time, he was a keen sportsman and his quiet enthusiasm in whatever he took in hand brought him into prominence in several directions. He had the gift of a charming affability, and because of his sound business acumen, his advice on commercial problems was sought frequently even on matters outside his particular business.

By his death beekeepers lost a friend and benefactor and one to whom the Somerset BKA owed a great deal for its flourishing condition at that time. The funeral took place at Canford Cemetery, and was attended by a large number of people beside the business staff, showing the great respect in which Col. Jolly was held. The Rev. W. S. Michell and Mr L. Bigg-Wither attended as representatives of SBKA.

H. J. Grist

Branch Secretary, visiting expert, queen raiser and then commercial beekeeper

One of the very early members, Mr Grist undertook the secretaryship of the newly formed Shepton Mallet District Branch and, with one other member, is listed in the year book for 1908. By the following year this branch had no fewer than 24 members. In the annual report for 1908, the Hon. Secretary, Mr Snelgrove states that *"as an illustration of his perseverance he has obtained the names and addresses of 60 other bee-keepers in his district whom it is hoped he may ultimately convince of the advantages of combination."*

In 1909 Mr Grist was successful in passing the Third Class Experts' certificate of the BBKA. His work for the local branch, both as secretary and visiting expert frequently received praise, but his key work was in connection with the re-stocking scheme after the Isle of Wight disease wiped out most colonies in

Mr H. J. Grist holding a Sladen queen cell carrier. *Photo: B.B.J.*

the county. He was one of four members selected to receive a share of the Dutch stocks imported in 1919 for breeding purposes and this was executed at his apiary at Evercreech (see page 34).

Mr Grist left Somerset in 1922, recommended by the Ministry of Agriculture, to become manager of the apiaries owned by Major Doreen Smith in the Isles of Scilly. He restored these from a derelict condition to a high state of efficiency. His next appointment was with Lord Pirrie, Whitley Park, Godalming, Surrey, for whom he established and managed apiaries consisting of over 300 colonies of bees. On the death of Lord Pirrie, his appointment was continued with Sir John Leigh, Bart, who had purchased the estate, lock, stock and barrel. It was at Whitley Park that Mr Grist died in February 1931.

Mr W. A. Withycombe

This is the story of a most remarkable man for whom beekeeping played a major part in life for nearly seventy years, and from which he largely gained his living. Born in Bath and brought up in Bridgwater, William Withycombe is recorded in the 1901 census as a 31 years old carpenter and bee expert. He started beekeeping at just eleven years of age and in his lifetime made a considerable contribution to the craft in Somerset and further afield during those unique years of transition from skep beekeeping to the movable comb hive.

On page 185 of Vol. 1 of Herrod-Hempsall's "Bee-Keeping, New and Old", can be read the following:

"In 1879, Mr. T. G. Newman, the then editor of 'The American Bee Journal,' visited his native town of Bridgwater, in Somerset. He talked bees to a youth named W. A. Withycombe, with the result that the latter became interested and made a couple of Langstroth hives and started beekeeping."

Mr Withycombe's apiary at Bridgwater, 1901. Left to right: Mr Withycombe's father, a helper, Mr Withycombe.

Mr Withycombe does not mention this encounter in connection with his commencement in the craft. He says that while still a boy at school he made a frame-hive from a pattern lent by a friend and commenced beekeeping with a swarm which promptly decamped the same day. A second swarm enjoyed the hospitality of the hive a little longer, remaining over one evening but taking flight the next morning. Not to be daunted, however, he later secured some driven bees to stock the hive, and after some lively times in which the fortunes of war fluctuated between the bees and the would-be beekeeper, he at last got the hive tenanted and the next season secured forty-seven sections. He built up to thirteen hives with the aid of stocks of bees, housed in straw skeps, which were obtained from a farm on the Minehead road beyond Cannington.

He studied what books were available, exhibited honey at shows and while still a teenager won three prizes at a Crystal Palace Show. He became one of the early members of the British Bee-keepers' Association and applied to take the "expert" examinations. To pass his examination for the 3rd class certificate he was obliged to travel from home to the Royal Show at Reading, Berkshire. It was the first time that a gift class had been tried and there were about 300 entries. Each of the candidates was asked by the judges whether they had won prizes for honey and each one who had done so was given a short trial and then set to judge batches of the honey. As his name was last on the list it was getting rather dark before his call came, but he managed to get through successfully, alas, missing the last train of the day. The indomitable spirit of this young man was such that he lightheartedly set off on foot arriving back at Bridgwater two days later. He was successful in the second class examination later the same season.

Early in the next year the examiner came to see him at Bridgwater. He asked him if he would go as expert for Kent and persuaded his father to let him go. While in Kent he was appointed as instructor at Swanley College and at Wye College, taking the place of the renowned Frank R. Cheshire who had recently died. During another season in Kent and Sussex Mr E. D. Till, a high-powered member of the BBKA Council and Kent BKA, authorised him to act as foul brood inspector. Should there be any disease in any members' hives he was to try to persuade them to burn them and Mr Till would give them a fresh stock free of charge in the autumn. This did not work very well, as only about one in three would comply and could not in those days be compelled to do so.

While in Kent he received an invitation from Sir John Lubbock (later Lord Avebury), to visit him at his residence near Beckenham and see his eight mahogany observatory hives and also his islands of ants. It was as a visiting expert that Mr Withycombe shone most brightly. In this capacity he served in several counties first in Kent and then Sussex, Essex, south Lancashire, Cheshire, Gloucestershire and finally back to his native county of Somerset.

In 1897 the old Bristol based Association consisted of ten districts and at this time its work was spreading very rapidly in Somerset. As a result of this another expert was required, and an advertisement was placed in the British Bee Journal. Mr Withycombe, a holder of the BBKA Expert Certificate, (second class), was appointed and he is recorded as the expert for Bridgwater and Ilminster districts. His work in this field is referred to elsewhere in this book.

Mr Withycombe was already well known among the beekeepers in parts of Somerset and he played a major role in the newly re-formed Somerest BKA being a member of the executive council from the beginning. It was said that some former members re-joined and many new ones were obtained through his influence. He visited them all twice a year, as he continued to do for forty years, using only a bicycle, and sometimes covering a thousand miles in one season. His practical skill and sound advice, coupled with unfailing benevolence and good humour endeared him to all who were helped by him. His lectures, sometimes illustrated by lantern slides, and his demonstrations, for many years given under the auspices of Somerset County Council Agricultural Instruction Committee, were always punctuated by little anecdotes concerning his adventures with bees and the odd things beekeepers sometimes do. Fearless and expeditious in handling bees, he sometimes said that he considered he had handled more bees and received more stings than any other man in England!

Mr Withycombe demonstrating at Seale Hayne College, Devon, circa 1930.

Over the years Mr Withycombe had tried every kind of hive available from the skep and Nutt's collateral to the National. He designed a hive which was excellent, but the record

W. A. Withycombe,

DOCKS.

BRIDGWATER,

Expert B.B.K.A.

Stocks, Swarms & Queens

Of specially selected hardy and prolific English Bees supplied.

Stocks from **15**/= each.   Swarms from **10**/= each.

Young, Fertile, Tested Queens, from June to August, **3/6** each.
September to November, **2/6** each.
Virgin Queens, **1/6** each.

All in self-introducing Cages.   Safe arrival guaranteed.

Driven Bees in August and September at **1/3** per lb.

W.A.W. has for several seasons been supplying some of the leading Queen Breeders with British Queens.

Advertisement in the Year Book for 1909

states that when cost was considered it was found to be outside the means of the ordinary cottager, unless a hire purchase system was adopted. He was highly adept at driving bees and had been known to cycle miles to save the lives of even a small stock. His involvement in beekeeping was wide ranging. He acted as demonstrator for the first ever beekeeping film. In another project he took an active part with F. W. L. Sladen in despatching bumble bee queens to New Zealand where they were required for the pollination of red clover.

In 1923 Mr Withycombe was a member of the newly formed sub-committee established to oversee the formation of experimental apiaries in the county to test new theories and methods. He took charge of the apiary at Cannington Farm Institute. In 1929 he became an official honey grader under the County scheme.

Having lived in Chilton Street by the Bridgwater docks for a great number of years, in 1928 the family moved to 32 Old Taunton Road. This was the central house of Lorne Terrace, built and lived in by his late father, John Withycombe and was home for the rest of his life. Nearby, at Rhode Lane he owned a piece of land on which he kept a large number of hives, and it was mainly from here that he ran his beekeeping operation. Apart from his expert's work he supplied beekeepers with stocks, swarms and queens of his specially selected hardy and prolific English bees and later in the year, especially in the early years, driven bees in August and September.

In 1945 Mr Withycombe was honoured by being elected a Vice-President of the British Beekeepers' Association. In Somerset he was on the SBKA Executive Council from the beginning and remained so until being made a Vice-President and Honorary Member at the end of the war. In 1946, as a result of the subscription fund, members of his Association had the pleasure of presenting him with an address and his portrait in oils in token of their esteem and affection, and in gratitude for his long, devoted and almost unrequited services over forty years. Mr Withycombe sat for the portrait in Mrs Colthurst's home at Wembdon. The artist was

Mr Withycombe shows bees to a friend, 1945.

Photo courtesy of Mrs P. Smith

Mr A. B. Connor of Burnham-on-Sea. It is a sizeable portrait and considered to be of high quality. No other member of this Association has, either previously or subsequently, been honoured in this way.

Mr Withycombe died on 1st February, 1949, aged 78. His memorial service held at King Street Methodist Church, Bridgwater, was attended by beekeepers from all over Somerset. Mr Snelgrove, President of the Association, paid the following tribute to him:

"As representing the British Beekeepers' Association and the Somerset Beekeepers' Association, of both of which our late friend, Mr. Withycombe was a distinguished member,

Connor's portrait of Mr Withycombe

I have been invited to pay a brief tribute to his memory on this mournful occasion.

Of his long life, nearly 70 years were devoted to the most fascinating, and at the same time the most difficult and elusive of all rural pursuits – the science and art of beekeeping. In this he worked mainly as a lecturer, and an expert visitor and adviser to those in difficulties. During the last 40 years, his activities were mainly confined to his native county of Somerset.

It is, I suppose, a human failing that we seldom fully appreciate our friends until we have lost them. It is given to but a few men to be valued at their true worth during their lifetime, but amongst that few our late friend surely held a high place.

No more shall we see his familiar figure setting out with bicycle and tools for a long trip over the Quantocks or Exmoor and returning at night, often wet and exhausted, only to set out again on the following day. No more shall we hear anxious beekeepers say: "Mr. Withycombe is coming next week and then all will be well."

Undaunted by bad weather, the steepest hills, or the fiercest bees, he never failed in giving much needed and often unrequited assistance to others. They marvelled at his vast knowledge, and admired his skill, but above all they loved to hear him talk of the wonders of bee life and his adventures amongst the bees.

Patient, courteous and kindly, ever ready to help and sympathise, modest in demeanour, yet firm in his convictions: speaking ill of no man, nor listening to it; not given to the pursuit of vain pleasures, but always ready to participate in innocent mirth, he was an exemplary man, a true friend, and – respected by all in his native town – a model citizen.

To him may well be ascribed the words of the great President Lincoln: "It is good for a man to be proud of his own city, and so to live that in the end his city may be proud of him."

The members of the Somerset Beekeepers' Association extend their deep sympathy to Mrs Withycombe and her son and trust that they will find consolation in the reflection that in his unselfish devotion to the service of others he realized one of the highest ideals in the life of a good man."

The following article, written by Mr Withycombe and published in the British Bee Journal of March 17, 1927, pages 116–117, paints a graphic account of his life, knowledge and interests as well as of country life at this time:

In the Verdant West

> "Where falls not hail, or rain, or any snow,
> Nor ever wind blows loudly: but it lies
> Deep meadow'd, happy, fair with orchard lawns
> And bowery hollows crown'd with summer sea." Tennyson

As a bearer of an old West Country name, and belonging to an old West Somerset family, I hope in this, and possibly, if welcome, succeeding letters, to interest your readers somewhat in this part of England, abounding as it does in ancient legends, customs and relics of ancient days, as well as present interest, even though its climate may not now quite come up to King Arthur's description. Having occasion to visit Porlock on February 1, I set out from Bridgwater by road, noticing before starting that my bees were flying freely and bringing in pollen of several shades, though the only flowers near were the laurustinus, winter heliotrope, and snowdrops, with a sprinkling of hazel catkins. Three miles out we reached Cannington with its County Agricultural College and an apiary of twelve stocks. In its early days the building was the home of the "Fair Rosamund," mistress of King Richard II, and on the hillside close by is the site of a furious battle against the Danes, who were defeated and their leader, Hubba, slain. Just beyond on the same hill is Brymore, the birthplace of John Pym, the Parliamentary leader, and in the grounds attached the gorse was flowering freely and bees were quite numerous on blossoms, while further on the winter heliotrope was growing in large patches by the roadside and scenting the air quite strongly for some distance. Near the seven milestone we pass a farm where I obtained my first stocks of bees, in straw skeps, but now it accommodates part of the apiary of our energetic Hon. Secretary, Mr W. West. Soon after we pass through Nether Stowey, beloved of the poets Coleridge, Southey, and Wordsworth. Near the ten milestone we can see the ruins of the abandoned copper mine which was worked for the Duke of Buckingham and Chandos until 1826.

Being on high ground we can see a large extent of the Welsh coast and have a fine view of the Steep Holm – the only place in England where the paeony grows wild – while a little to the right we can see the range of the Mendip Hills and Cheddar Gorge, the home of another wild flower, the Cheddar Pink. Glastonbury and its Tor, the home of the Holy Thorn, which blossoms at Christmas time, can also be seen. On the near shore of the Bristol Channel is Skurton Bars where every autumn at St. Matthew's flood tide the conger eels are hunted with dogs and handspikes, an exciting sport which I have enjoyed on several occasions. Close by is Kilton church, which still keeps its barrel organ, which has a remarkably sweet tone. Passing on through the woods we come out on an open expanse of gorse, heather, and whortleberries, which extends along the Quantock Hills for several miles, and is a grand foraging ground for bees, many being kept at Holford by one large and several smaller apiarists. After a sharp descent at Kilve, twelve miles, the road is nearly all uphill until we reach St. Audries Park, now cleared of its deer, and being rapidly cleared of its trees also, to the great detriment of its former beauty. Almost on the edge of the park lives a bee-keeper who has been experimenting with *Braula coeca* as a cure, or consumer of, the *Acarapis woodi*. A long descent now takes us into Williton, with many of its cottage walls ornamented by slabs of alabaster from the cliffs at Watchet. Here lives Dr Killick, Vice-Chairman, SBKA, where we have been privileged to spend many interesting and instructive hours in examining the bees and studying a small part of his extensive collection of microscopic slides. Close by is Orchard Wyndham, the ancient seat of the Wyndham family, its present owner being one of our Vice-Presidents. At Washford, twenty miles, we turn to the left and pass the remains of Cleeve Abbey on our way to the Holy Well and Chapel of St. Pancras, now owned and occupied by the genial Hon. Secretary and expert of the Minehead branch, SBKA, who is a large and successful bee-keeper, and after enjoying his hospitality and admiring the ingenious way in which he has adapted the Holy Well to automatically deliver water to his house, we go out and inspect his long line of seventy hives, which are located in a very fertile-looking valley, wonderfully well sheltered, and quite accessible from his house. Bees here were flying freely, and there seemed to be a fair amount of forage available, with a promise of abundance in its proper season. On leaving here I could not help wondering whether the ancient monks had created a kind of corner in holy water, for the name of the adjoining parish is Monksilver. Reaching the main road again, at about two miles distance, we see a signpost with a familiar name, Withycombe, and can just make out the tower of the whitewashed church on our left, while much nearer is a field of mustard nearly in full bloom. Half-a-mile on we come to Carhampton, the home of many legends, many old customs, and the real home of the Withycombe family, for the churchyard, which is the largest in Somerset, contains the tombs of many generations, dating back for the last three centuries. As for the old customs, I quote one from the "West Somerset Free Press" of January 22 last:

"A merry band set out from Carhampton on Monday night, the eve of old twelfth night, to wassail the apple trees. It is recorded that on one occasion the farmers had been so hospitable that when they came to the last farm on the list, the revellers solemnly wassailed a lilac bush in mistake for an apple tree. However, nothing like this occurred at Carhampton on this occasion, for they cheerfully gathered around the largest apple tree in Mr Tarr's orchard which overlooks the churchyard and Blue Anchor Bay. A large bucket of cider was provided, into which the chief wassailers dipped mugs and handed them round. Mr Squance then struck up the wassail song, which was taken up by those present as follows:

'Old apple tree, old apple tree,
We've come to wassail thee,
Hoping thou us will bear,
For the Lord doth know where
We shall be, to be merry next year.
To bloom, to bear – so merry let us be;
Let every man take up his cup,
And health to the old apple tree.
Here's to thee, old apple tree,
Mayst thou grow apples enow,
Hats full, caps full, three bushel bags full.
Big barn floors full, and a little heap under the stairs.
 Hip, hip, hurrah!'

"The wassailers then cheered, shouted and sang, while guns were fired into the air, and pandemonium was let loose. The robins were not forgotten: pieces of toast dipped in cider being placed in the forks of the trees for their consumption."

The chief legend of Carhampton concerns the famous round table of King Arthur, which is said to have been given to St. Carantoc by an angel, and then floated down from the Upper Severn to Blue Anchor Bay, where King Arthur was hunting, and the table was there presented to the King by the Saint. Carhampton Church is well worth a visit, for it has a beautiful altar screen and an enormous Peter's Pence box. Two miles on is Dunster, with its ancient castle, the home of the Luttrell family and as we pass on the road toward Minehead, the fresh green of the new shoots of feverfew and fool's parsley by the roadside shows that spring is at hand. Just before reaching Minehead half-a-dozen stocks of bees in well-kept W.BC. hives can be seen on the right, and, just after leaving, there is another apiary of twelve hives, also on the right-hand side of the road. Each owner did remarkably well with his bees last season. No more bees are seen till we reach the smithy at Allerford, where three stocks are kept. Then a visit is paid to Selworthy rectory, from which one of the most delightful views in the county can be obtained, and the picturesque cottages at the green are also worth seeing. On toward Porlock, soon after crossing the Horner stream, another field of mustard is seen in flower, and in Porlock itself roses and marguerites are in bloom. Mr Smith reports having taken over 1,000lb of honey last season, and a friend whom he assists secured 600lb. Bees were flying freely and bringing in pollen, especially a Carniolan stock of Mr Smith's.

Our outward journey ends at Porlock Weir, passing on the way the house and apiary of Mr Cook, an expert of the association, and here in the open ground there is quite a wealth of bloom. Though the 30ft. fuchsias have all been cut down and removed, perhaps for firewood, there are hedges of hydrangeas and veronicas of many kinds, roses, geraniums, stocks, wallflowers, pansies and violas, daffodils in abundance, all flowering freely along with many other flowers, and the waste land at the roadside is crowded with winter heliotrope, the heavy fragrance of this again greeting us here and there on our return journey of thirty-four miles.

W.A. Withycombe
Docks, Bridgwater

Lovelace Bigg-Wither, Honorary Secretary, 1912 – 1922

Lovelace Bigg-Wither was the only son of Arthur Fitzwalter Bigg-Wither and grandson of the Rev. Lovelace Bigg-Wither, landed gentry of Manydown Park, near Wootton St. Lawrence in Hampshire. Among his ancestors was Harris Bigg-Wither whose proposal of marriage was accepted and then rejected by Jane Austen. He appears to have had business interests, and like Cowan, had ample time to follow leisure pursuits. These were horticulture, beekeeping, natural history, archaeology and aviation. From 1910 until his death he was Hon. Secretary of the Mendip Nature Reserve Committee.

After a successful school and college career, at the age of eighteen, Bigg-Wither went to India where for six years he was a tea planter in the Darjeeling district. On returning to England he studied horticulture and beekeeping at Swanley Horticultural College. The beekeeping instructor was William Herrod-Hempsall. After some years in Kent, in 1907 he and his wife, Muriel set up home at Birdwood, Wells and he joined the Somerset BKA around 1909. Whatever the reason that brought Mr Lovelace Bigg-Wither to Somerset, it was indeed the Association's good fortune. He was soon actively involved and in 1912, after a short spell of assisting Mr Snelgrove, succeeded him as Honorary Secretary. This post he held

Mr L. Bigg-Wither after co-judging honey at Taunton Flower Show, 1935

valiantly for a ten year period embracing not only the ravages and effects of World War I but those dreadful years of the I.O.W. epidemic. The latter had dire effects on the bee population and membership. He organised and brought to a successful conclusion the re-stocking scheme following the severe losses incurred.

In 1922 he moved to the Vicarage at Cleeve, near Bristol, continuing his work as divisional Secretary of the Central Division and Secretary of the Wells Branch as well as being the visiting expert. A member of the Somerset BKA Council and delegate to the BBKA, he also, while living at Cleeve, supervised the Association's experimental apiary at Long Ashton Research Station. Mr Bigg-Wither already held the BBKA's Expert certificate (final examination) and became an examiner himself. He was, for a great number of years, one of two lecturers in beekeeping for the Somerset County Council, the other being Mr Withycombe. He was a well known honey judge and had officiated at every major show in the country.

In 1925 the Bigg-Withers returned to Wells, where the beautiful garden at Colmar's Ash was testimony to his skill and success as a horticulturist. He took a great interest and an active part in the Association until the end of his life.

Mr Bigg-Wither's Apiary at Wells.

On 4th November 1938 Mr Bigg-Wither, in apparent good health, judged at the Bath Show. The following day he became ill and died in Wells Cottage Hospital on 18th November at the age of 63. His funeral was at the Church of St. Thomas of Canterbury, Worting, west of Basingstoke, where he was interred in the family vault. A Vice-President of the Association since 1924, he was a member for almost thirty years. He is remembered through the Association's Bigg-Wither Library of which some of his beekeeping books formed the foundation. At the Mid Somerset Show the Bigg-Wither Challenge Cup is competed for annually by members of the Central Division.

Mrs Bigg-Wither, who had passed the BBKA Expert (2nd class) examination in 1920, shared her husband's interest in beekeeping. She continued for many years as President and visiting expert of the Wells Branch.

Harold H. Meade

Somerset born and bred, Harold Meade spent his early years in Middlezoy. As a boy he had a passion for helping on the local barges that plied between Bridgwater and Curload.

Some prominent Members

These particularly slow, quiet and restful journeys along waterways bordered by a wide range of plants, alive with birds and other forms of wildlife made him what he was – a quietly spoken countryman with a keen eye to observe all that was happening around him. A knowledgeable ornithologist and keen eel trapper, Harold had many other interests and activities and was a good shot.

His early years were spent coopering and basket-making. During World War II the Ministry of Defence found that willow grown on the levels of Somerset was the best material to withstand the shock of parachuting. Baskets made of willow were used to drop supplies to troops in difficult situations. Harold was engaged in vital work concerning this.

Harold Meade, holding a fledgling owl.
Photo courtesy of Mr R. H. Meade

After the war, when Harold owned withy beds, he co-operated with the Long Ashton Research Station in combating the various diseases besetting the willow. Experts engaged on this important work were based at the King's Head, Athelney, owned and run by Mr and Mrs Meade.

This public house would have pleased King Alfred, for Harold kept it a secret from strangers. There was no sign. Two ciders were served, and also mead, all made on the premises. Locals kept their own jugs in the parlour, helped themselves as required, putting their money in a drawer and taking the change without assistance.

It was Harold's mother who first interested him in bees. She told him her memories of how, at the end of the summer, the bees were killed using sulphur. She explained to him how the honey was strained, the crushed combs soaked in rain water and then fermented with yeast applied on pieces of toast to make mead.

This early interest in beekeeping became Harold's great love and he commenced in the 1930s. He collected a swarm from an allotment and kept them in a box at home. After taking the honey the bees died because he did not understand about autumn feeding. The following year Harold heard of a swarm at New Bridge so he set off by motorcycle, took the swarm, and transported it to Athelney in a box strapped over his shoulders. When he reached home there was not a bee remaining! Not to be thwarted, he then purchased four stocked hives, immediately selling one to someone else. This person arrived to collect it complete with several jars, assuming that the hive would be laden with honey. Harold joined the Association around 1936 and was then tutored by the well known expert, Mr Withycombe. In turn he too helped many beginners over the years. The early outdoor meetings of the Bridgwater Division were held at Athelney during dandelion flowering time. Harold always coaxed his colonies to be strong early for this important harvest.

Despite his secret location, producers for both radio and television managed to find him over the years, and included him in their various programmes about country life.

Harold Meade was a Vice-President of the Association and President of the Bridgwater Division until his death in 1984, aged 83.

Chapter 12: Miscellany

The Association's Logotype

The earliest use of this distinctive logo is on the front cover of the Annual Report, List of Members and Balance Sheet for 1924. The skeps motif is clearly seen on the official honey label of the Association illustrated on page 142, as early as 1909, but it is a different impression and does not bear the Latin inscription around it.

As you will read in Mr Snelgrove's profile, in 1922, resulting from his research in Latin literature relating to Roman Bee Lore, he was awarded a Masters' degree. There can be no doubt that Mr Snelgrove was the creator and designer of this logotype. The inscription, however – "*Hiemis memores aestate laborem experiuntur*" – is not of his authorship, though we must credit him for his appropriate choice. The words are taken from Virgil's Georgics, Book IV, a book he certainly would have studied. He has selected part of a long sentence that in the translation by Cecil Day Lewis reads thus:

"*Aware that winter is coming, they use the summer days for work and put their winnings into a common pool.*"

There are numerous translations of the Georgics. In English there are translations by Cowper, Dryden, Mackail, Martyn and Pope to name but a few, each reflecting contemporary English usage. An older and rather more florid version is as follows:

"*They alone know a fatherland and fixed home, and in summer, mindful of the winter to come, they spend toilsome days and garner their gains into a common store.*"

Another writer's poetic form reads thus:

"*To home and country they alone are true,
And mindful of the winter soon to come,
Work hard in summer, to the common store
Contributing their gains.*"

Honey Labels and Seals

One of the first services offered to members after the Association's inception was the provision of an official honey label and guarantee slip, the latter also serving as a

The Association's original design of Honey Label and Guarantee Slip

Miscellany

tamper-proof seal. The early year books all contained sample labels and slips as illustrated, together with an application form for purchasing them. The cost at that time was one shilling and threepence (6p) per 100, postage free. The label is illustrated opposite.

At some time between the wars the label design was changed and made available, not only in a choice of black or blue (an advancement on Ford motor cars!), but for both 1lb and ½lb jars and honeycomb sections.

In 1929 a great step forward was made, mainly through the endeavours of the President, Mr Snelgrove, by the introduction of a honey grading scheme. Somerset BKA was the first in the country to inaugurate such a scheme with full approval and assistance from the Ministry of Agriculture and Fisheries, the Ministry paying almost the whole cost of the initial outlay.

In the first year 83 samples of honey from members were submitted and the results of the grading were as follows:

Samples	Grade	Weight (tons cwts. lbs.)	Labels issued
12	A1 representing	1 6 2	3,200
61	A "	5 6 0	12,120
10	B "	5 2	900
83		Tons 6 18 4	16,220

The Assistant Hon. Secretary, Richard Beck administered the scheme. The judges appointed were Messrs Withycombe and Rudman, but later there was a grading committee. The scheme proved popular with members for many years. By 1932 grades A1 and B appear to have been dispensed with, but of nineteen samples submitted in that year, only one was disqualified and 2,500 labels were issued. At one stage the regulations were very strict. Honey samples had to be in jars at least five inches high so that the density test could be applied.

In later years there was a grader for each Division. The rules governing the scheme were revised several times, but the 1972 revision made it clear that the producer completely indemnified the Association against any action arising from their use. By 1978 supplies of the labels and accompanying seals were exhausted. Due to the prohibitive cost of reprinting, and the fact that only a few members availed themselves of the scheme Council decided to discontinue it. The illustrations on the previous page show the last version of the label, which was green and gold, and the accompanying seal after the die had been altered to include the weight in metric units.

Incentives and Rewards for competitive Classes at Shows

Medals

In earlier times, particularly before Associations owned many cups or other trophies it was usual, where appropriate, to present medals in silver or bronze for excellence. The example shown was won by Mr R. Addision of North Petherton in 1897 and was recently sold by auction over the Internet for £28. Mr Addision was a member of Bridgwater and District Division until his death in 1940. The medal is solid silver, not hall marked, 45mm in diameter and about 4mm in thickness. The obverse has a beautifully crafted embossed design on it, with daisies, roses, fuchsias, etc. around a skep and wooden hive design. Beneath is a scroll engraved "Henbury 1897". The reverse is inscribed "Bristol, Somersetshire & South Gloucestershire Bee-Keepers' Association – First Prize: Mr. R. Addison, North Petherton for best 6 sections of honey.

Cups and other Awards

These are all "perpetual" and are held only for one year, being returned to the organisers in time for the following year's show.

County Awards

Jubilee Cup, Miss M.D. Bindley Cup, Terry Arnold Challenge Trophy, Duffin Challenge Trophy, Clifford Langford Award.

County Cups.   Photo: A. D. Charles

Miscellany | 45

Central Division Awards

Allen Challenge Cup, Bigg-Wither Challenge Cup, Wells Branch Challenge Trophy.

Mendip Division Awards

Downs Cup, Dymboro Cup, Gulliford Cup, Haydon Cup, Lye Cup, Weeks Cup.

Mendip cups. *Photo: B. J. Newton*

Somerton and District Division Awards

Coronation Challenge Cup, Morris Plaque, Alan Bromley Cup, Bill Harris Cup, John Lindars Memorial Trophy, Perkins Wine Goblet, C. H. G. Langford Cup, Somerton Divisional Trophy.

Taunton and District Division Awards

Downes Cup, Taunton Challenge Trophy, Priscott Skep Plate, John Spiller Mascot, Pat Barter Silver Cup, Stoker Challenge Trophy, Novice Trophy, Taunton Show Trophy.

Three cups, belonging to the former Bridgwater and District Division: the W.A. Withycombe Memorial Cup, The Beatrice Turner Memorial Cup and the Douglas Henry Spurrell Memorial Cup have been missing for some years.

Taunton trophies. *Photo: A. D. Charles*

Bristol Branch Trophies

Although the Bristol beekeepers are no longer part of SBKA details of two trophies are given for historical reasons and because of their exceptional interest.

The Bristol Silver Queen. Photo: A. D. Charles

Bristol Silver Queen: This trophy was donated anonymously to the Bristol Branch and was first awarded at the second Bristol Honey Show held in 1929. The honey classes were judged by W. Herrod-Hempsall and the photograph on page 43 was probably taken on that occasion. Initially exhibitors from many parts of the country sent exhibits to Bristol in an attempt to gain this prestigious trophy for one year and for the accompanying monetary prize. Between 1948 and 1953 the Silver Queen was awarded at the Bath Honey Show. It was said that this came about due to lack of suitable venues in Bristol after wartime bombardment. It can just be distinguished in the photograph on page 51. The cup is currently awarded at the Bristol Flower Show held on Durdham Down each August.

Originally this trophy was awarded to the exhibitor of the best 4 sections and 4 jars of run honey. Nowadays it is awarded to the exhibitor gaining the highest points in the open classes. The trophy has been won by the following SBKA members:

1963: Bert and Len Gulliford of Clapton, Midsomer Norton, (Eastern Division)

1964: Jim Hannam of Wells and West Country Honey Farms, (Central Division)

1967, 1969, 1970: Stan Burchill, Porlock, (Exmoor Division but formerly of Bristol)

1971, 1972: David Morris, Halse, (Taunton Division).

Snelgrove Cup: This trophy was presented to the Bristol Branch by Mr L. E. Snelgrove in 1935 for the best exhibit at the Bristol Honey Show by a Bristol member and was first won by Mr E. Lewis of Abbots Leigh. It is a good sized silver cup with a lid and finial, all in art deco style. In 1938 Mr G. H. Jenner of Chard who later became CBI for Devon won the trophy. This is curious because it is unlikely that he was a Bristol member. The war probably caused the Bristol Branch to lose track of the whereabouts of this cup, but it was missing for almost fifty years from that time. Mr Jenner may have been under the impression that he had won it outright. The cup was spotted by a vigilant beekeeper on Mrs Jenner's mantelpiece after the death of her husband. It was returned to Bristol and has been awarded each year since 1986.

Presidential Chain of Office

At a Council meeting held on 12th September 1992, the Chairman, Mr J. C. Newcombe floated the idea of acquiring a chain of office for the president to wear on appropriate occasions and it was agreed to consider this. At the December meeting some delegates expressed the view that members may resent the Association bearing the cost from funds.

The matter was quickly resolved, however, by a generous offer from the Hon. Treasurer, Mr Roland Dell, to donate a suitable chain. Furthermore, he agreed that bars would be inscribed with the names and dates of previous presidents of the Somerset Beekeepers' Association from the time of its inception in 1906. Successive presidents have worn this chain of office since February 1993.

West Country Honey Farms Ltd. Award

After Mr J. Hannam sold West Country Honey Farms and retired to Tenerife the new owners felt that they too would like to establish a close link with the Association. Following negotiations between representatives of the Association and the Factory Manager, Mr P. Warren, terms were agreed and the Somerset Beekeepers' Association and West Country Honey Farms Challenge Rose Bowl Competition was inaugurated in 1991. The award would be made annually for excellence in apiculture, the winner to be decided by a panel of three expert adjudicators, one to be from outside the county. The recipient would not only hold the Rose Bowl for one year but also receive a complete National hive.

The West Country Honey Farms Rose Bowl.
Photo: A. D. Charles

In December 1992 a letter was received from Mr Warren informing the Association that tight financial control on all overheads meant that the firm would not be continuing with the competition. The Rose Bowl would become the property of the Somerset BKA. Subsequently it was decided that the rules for adjudication were too difficult and expensive to administer so that the award would no longer be for excellence in apiculture.

From 1995 the Rose Bowl has been awarded annually to the member who is adjudged to have contributed most to beekeeping and the SBKA in the last calendar year.

Recipients to date are as follows:

1991	*S. N. Jones	1998	W. W. M. Milton
1992	*S. J. Gammon	1999	A. C. Elcomb
1993	*T. B. Trood	2000	N. B. Trood
1994	*S. Blake	2001	E. Currell
1995	P. J. Rich, N.D.B.	2002	S. Perkins
1996	G. Fisher	2003	K. A. T. Edwards
1997	J. C. Newcombe	2004	A. D. Charles

* Awarded under the initial criteria

Internal Communication

Nobody would disagree that communication with members is an essential ingredient of running a successful association. During the last hundred years communication has taken various forms with varying degrees of frequency and effectiveness. The main modes of communication over the years have been as follows:

The Year Book

This publication has been produced annually since our inception except in 1989 and 1990 Every member is entitled to a copy. It is mainly a reference book and early issues contained the following information:

1. A list of officers of the association and of its Divisions and Branches
2. A list of the members of the association by Divisions and Branches
3. The rules of the Association
4. Privileges of membership
5. The balance sheet
6. Application forms and samples of the Association's honey labels and seals.
7. A calendar of beekeeping work
8. Some useful hints on beekeeping.

Over the years efforts have been made to increase its value as a database and to broaden its interest. At various times there have been regular features such as the Crop and Weather Report for the county and statistics regarding the incidence of diseases. Articles on topical matters by experts also appeared, mostly from authors within the membership.

The visiting Expert

Much has already been said about the visiting experts. Until about 1945 members were entitled to one or two visits per annum from an association expert, who, apart from assisting and advising on matters of husbandry was a vital link with the association and the craft. The expert would bring snippets of news, collect subscriptions if due, for onward passage to the treasurer, and keep members in touch with topical matters of beekeeping. Most beekeepers at one time regarded this as the most valuable benefit of membership. After the appointment of a full time beekeeping instructor in 1944 expert work gradually fell into decline. By the early 1950s they were referred to as advisory officers and the entitlement to visits was on a different basis. Presently these advisory officers, where divisions appoint them, play a less significant role.

Notices and News Sheets

In the early days the GPO provided a speedy and reliable postal service with two deliveries a day in most areas. The cheapest way of using the postal service for short messages was the ubiquitous postcard. These could be printed cheaply in bulk and if desired, with spaces left for local secretaries to insert by hand the dates, times and venues for local meetings.

> **STREET & GLASTONBURY BEE-KEEPERS' ASSOCIATION.**
>
> THE ANNUAL MEETING will be held in the garden at PORTWAY HOUSE, STREET, on WEDNESDAY, AUGUST 8th. 1923.
>
> 4.0 p.m.—Demonstration.
> 5.0 „ —Pic-nic tea on the lawn. All attending are requested to bring their own eatables. Cups of tea will be provided.
> 5.30 „ —Business Meeting.
> 6.0 „ —Bee Talk by RICHARD BECK, (of Clevedon). The I.O.W. Disease Mite, etc., will be shown through microscopes.
>
> *please invite anyone interested* EDWIN I. WALKER, Hon. Sec.

As the 20th century progressed there were great advances in methods of producing large quantities of documents. One common method in use for a great number of years involved cutting a stencil from a special skin of waxed paper using the typewriter keys. This fitted on to the roller of a rather hefty office machine, was inked, which could be a messy operation, and hundreds, if necessary, of copies could be "rolled off" in a short time. This process was useful, not only for the production of news sheets but also for circulating minutes of meetings, balance sheets, etc. Committee members could then have their own copies in advance instead of waiting until the next meeting for minutes of the previous meeting to be read. Many local secretaries who were teachers or did office work had access to such machines, but some associations owned one.

We share with Devon Beekeepers' Association

Devon and Somerset had long been closely associated in beekeeping matters, having from 1935 to 1947 been the two associations forming the major part of the South-Western Beekeepers' Federation. Somerset BKA had long desired to participate in the advantages of sharing in the Devon BKA official organ entitled "Beekeeping" that was originally the Federation journal.

Captain Stoker, a former Hon. Secretary and Mr G. M. Bell of Yeovil negotiated that a portion of each issue would be reserved for Somerset news and contributions, Mr Bell acting as sub-editor for Somerset. We were allotted four sides. This arrangement lasted from 1949 until 1954. During this time there were regular contributions from Mrs Colthurst who wrote interesting and detailed crop and weather reports. Sometimes there was an article or account from Mr Baker (Zummerzet) and accounts from Divisions and Branches about their apiary meetings. There were accounts of the main meetings and rallies. They mostly related to events that had already occurred, rather than detailing anything that was to come. Towards the end of this time most pens had dried up and there was much less divisional news, but Mrs Colthurst wrote as prolifically as ever until the scheme finished.

By the 1980s the practice of letter writing became less as most people, by then, possessed telephones. The snag with making arrangements or passing messages by telephone is that people fail to take notes, make mistakes with them, or simply forget a key point of the conversation. With no material record, mistakes could then occur, such as turning up for a meeting on the wrong day or at the wrong time.

A regular Newsletter

In 1977 David Charles persuaded the SBKA Council that a regular newsletter, in addition to the year book, was needed in order to improve contact with the membership. This was approved and David's friend and fellow member, Peter Goolden of Glastonbury took on the editorship for a short time, being succeeded by Mr J. Fieldhouse of Enmore and a number of other people over the years. David Charles produced one or two issues between editors to keep it going. Mrs Joyce Bromley of Aller, a member of the Somerton and District Division took it on for a time in the 1980s and it was an advance that she typed the copy into a computer. David then took a master copy to St. Andrews Press at Wells and sat watching over 500 copies pour from their modern Xerox machine in a matter of minutes. Michael Milton followed Joyce Bromley as editor ably fulfilling the role for several years. In 2000 Steve Wawman of Exmoor Division took over the task until 2005 when the young Richard Bache of Somerton and District Division succeeded him.

The Computer Age arrives

The biggest advance in communications has to be the development and widespread use of the personal computer. By the millennium most beekeepers seemed to have one and the practice of sending notices of committee meetings, agendas and minutes by electronic means was becoming standard practice. Members could opt to have their newsletters sent electronically instead of through the post.

The Internet

In the spring of the year 2000 the Association launched its first website. The instigator of this was Gerald Fisher, the Association's Chairman at this time. He devised much of the content and with help was able to design and launch it. The website contained details about the Association, its branches, officers, and programmes of events. In addition it carried

Miscellany

information about bees and beekeeping directed towards anyone who chose to look at the site. This has been beneficial in attracting interest from outside the Association, in recruitment and enrolling people for introductory classes.

In 2005 the website was completely re-designed and upgraded. This new site was the brainchild of Steve Horgan, Gerald Fisher's son-in-law. Steve designed the site and Gerald compiled the contents. It was entered in the website class of the competitive section at the 39th Apimondia International Apicultural Congress held in Dublin during August of this year, attended by several Somerset members. Of the twenty-six entries received for competition the adjudicators considered the Somerset entry to be the best and awarded a gold medal for it. As Gerald did not attend the Congress the Somerset members present agreed among themselves that David Morris, the Association's Publicity Officer should be the representative to accept this medal on behalf of the Association. During the closing ceremony he was formally presented with it by Mr Dermot Ahearn, Ireland's Minister for Foreign Affairs. The Association's international success received wide press publicity.

The website address is http://www.somersetbeekeepers.org.uk

Apimondia Gold Medal 2005, obverse.
Photo: R. A. Bache

Apimondia Gold Medal 2005, reverse.
Photo: R. A. Bache

Appendix 1: Principal Officers

Year	President	Chairman	Secretary	Treasurer
1906	Lady Smyth	Lt.Col. H.F. Jolly	Mr L.E. Snelgrove	Mr L.E. Snelgrove
1907	Lady Smyth	Lt.Col. H.F. Jolly	Mr L.E. Snelgrove	Mr L.E. Snelgrove
1908	Lady Smyth	Lt.Col. H.F. Jolly	Mr L.E. Snelgrove	Mr L.E. Snelgrove
1909	Lady Smyth	Lt.Col. H.F. Jolly	Mr L.E. Snelgrove	Mr L.E. Snelgrove
1910	Lady Smyth	Lt.Col. H.F. Jolly	Mr L.E. Snelgrove	Mr W.A. Withycombe
1911	Lady Smyth	Lt.Col. H.F. Jolly	Mr L.E. Snelgrove	Mr W.A. Withycombe
1912	Lady Smyth	Lt.Col. H.F. Jolly	Mr L. Bigg-Wither	Mr L. Bigg-Wither
1913	Lady Smyth	Lt.Col. H.F. Jolly	Mr L. Bigg-Wither	Mr L. Bigg-Wither
1914	Lady Smyth	Lt.Col. H.F. Jolly	Mr L. Bigg-Wither	Mr L. Bigg-Wither
1915	Mr T.W. Cowan	Lt.Col. H.F. Jolly	Mr L. Bigg-Wither	Mr L. Bigg-Wither
1916	Mr T.W. Cowan	Lt.Col. H.F. Jolly	Mr L. Bigg-Wither	Mr L. Bigg-Wither
1917	Mr T.W. Cowan	Lt.Col. H.F. Jolly	Mr L. Bigg-Wither	Mr L. Bigg-Wither
1918	Mr T.W. Cowan	Lt.Col. H.F. Jolly	Mr L. Bigg-Wither	Mr L. Bigg-Wither
1919	Mr T.W. Cowan	Lt.Col. H.F. Jolly	Mr L. Bigg-Wither	Mr L. Bigg-Wither
1920	Mr T.W. Cowan	Lt.Col. H.F. Jolly	Mr L. Bigg-Wither	Mr L. Bigg-Wither
1921	Mr T.W. Cowan	Lt.Col. H.F. Jolly	Cdr R.E. Graham, RN	Cdr R.E. Graham, RN
1922	Mr T.W. Cowan	Mr L.E. Snelgrove	Cdr R.E. Graham, RN	Cdr R.E. Graham, RN
1923	Mr T.W. Cowan	Mr L.E. Snelgrove	Cdr R.E. Graham, RN	Cdr R.E. Graham, RN
1924	Mr T.W. Cowan	Mr L.E. Snelgrove	Mr William West	Mr William West
1925	Mr T.W. Cowan	Mr L.E. Snelgrove	Mr William West	Mr William West
1926	Mr T.W. Cowan	Mr L.E. Snelgrove	Mr William West	Mr William West
1927	Mr L.E. Snelgrove	Dr J. Wallace	Mr William West	Mr William West
1928	Mr L.E. Snelgrove	Dr J. Wallace	Mr William West	Mr William West
1929	Mr L.E. Snelgrove	Dr J. Wallace	Mr William West	Mr William West
1930	Mr L.E. Snelgrove	Dr J. Wallace	Mr William West	Mr William West
1931	Mr L.E. Snelgrove	Dr J. Wallace	Mr William West	Mr William West
1932	Mr L.E. Snelgrove	Dr J. Wallace	Mr William West	Mr William West
1933	Mr L.E. Snelgrove	Dr J. Wallace	Mr William West	Mr William West
1934	Mr L.E. Snelgrove	Dr J. Wallace	Mr William West	Mr William West
1935	Mr L.E. Snelgrove	Dr J. Wallace	Mr William West	Mr William West
1936	Mr L.E. Snelgrove	Dr J. Wallace	Mr William West	Mr William West
1937	Mr L.E. Snelgrove	Dr J. Wallace	Mr William West	Mr William West

Principal Officers

1938	Mr L.E. Snelgrove	Dr J. Wallace	Mr William West	Mr William West
1939	Mr L.E. Snelgrove	Dr J. Wallace	Mr L.N. Pardoe	Mr M. Barnett
1940	Mr L.E. Snelgrove	Dr J. Wallace	Mr L.N. Pardoe	Mr M. Barnett
1941	Mr L.E. Snelgrove	Dr J. Wallace	Mr K.T. Batty	Mr K.T. Batty
1942	Mr L.E. Snelgrove	Dr J. Wallace	Mr K.T. Batty	Mr K.T. Batty
1943	Mr L.E. Snelgrove	Dr J. Wallace	Mr K.T. Batty	Mr K.T. Batty
1944	Mr L.E. Snelgrove	Dr J. Wallace	Capt. T.G. Stoker	Capt. T.G. Stoker
1945	Mr L.E. Snelgrove	Dr J. Wallace	Capt. T.G. Stoker	Capt. T.G. Stoker
1946	Mr L.E. Snelgrove	Dr J. Wallace	Capt. T.G. Stoker	Capt. T.G. Stoker
1947	Mr L.E. Snelgrove	Major J. Lintorn Shore	Capt. T.G. Stoker	Capt. T.G. Stoker
1948	Mr L.E. Snelgrove	Mr L. Hender	Capt. T.G. Stoker	Capt. T.G. Stoker
1949	Mr L.E. Snelgrove	Mr L. Hender	Mr E.J.F. Betts	Mr E.J.F. Betts
1950	Mr L.E. Snelgrove	Mr L. Hender	Mr E.J.F. Betts	Mr E.J.F. Betts
1951	Mr L.E. Snelgrove	Mr L. Hender	Mr E.J.F. Betts	Mr E.J.F. Betts
1952	Mr L.E. Snelgrove	Mr L. Hender	Mr E.J.F. Betts	Mr E.J.F. Betts
1953	Mr L.E. Snelgrove	Mr L. Hender	Mr E.J.F. Betts	Mr E.J.F. Betts
1954	Mr L.E. Snelgrove	Miss M.D. Bindley	Mr E.J.F. Betts	Mr E.J.F. Betts
1955	Mr L.E. Snelgrove	Miss M.D. Bindley	Mr E.J.F. Betts	Mr E.J.F. Betts
1956	Mr L.E. Snelgrove	Miss M.D. Bindley	Mr E.J.F. Betts	Mr E.J.F. Betts
1957	Mr L.E. Snelgrove	Mr L. Hender	Mr E.J.F. Betts	Mr E.J.F. Betts
1958	Mr L.E. Snelgrove	Mr L. Hender	Mr E.J.F. Betts	Mr E.J.F. Betts
1959	Mr L.E. Snelgrove	Mr L. Hender	Mr E.J.F Betts	Mr E.J.F. Betts
1960	Mr L.E. Snelgrove	Mr L. Hender	Mr E.J.F. Betts	Mr E.J.F. Betts
1961	Mr L.E. Snelgrove	Mr L. Hender	Mr E.J.F. Betts	Mr E.J.F. Betts
1962	Mr L.E. Snelgrove	Mr L. Hender	Mr E.J.F. Betts	Mr E.J.F. Betts
1963	Mr L.E. Snelgrove	Mr L. Hender	Mr J.C. Reynolds	Mr J.C. Reynolds
1964	Mr L.E. Snelgrove	Mr L. Hender	Mr J.C. Reynolds	Mr J.C. Reynolds
1965	Mr L.E. Snelgrove	Mr L. Hender	Mr J.C. Reynolds	Mr J.C. Reynolds
1966	Mr L. Hender	Mr R.W. Sawyer	Mr J.C. Reynolds	Mr J.C. Reynolds
1967	Mr L. Hender	Mr R.W. Sawyer	Mr J.C. Reynolds	Mr J.C. Reynolds
1968	Mr L. Hender	Mr R.W. Sawyer	Mr J.C. Reynolds	Mr J.C. Reynolds
1969	Mr L. Hender	Mr R.W. Sawyer	Mr J.C. Reynolds	Mr J.C. Reynolds
1970	Mr L. Hender	Mr R.W. Sawyer	Mrs R.E. Lovegrove	Mrs R.E. Lovegrove
1971	Miss M.D. Bindley	Mr R.W. Sawyer	Mrs R.E. Lovegrove	Mrs R.E. Lovegrove
1972	Miss M.D. Bindley	Mr R.W. Sawyer	Mrs R.E. Lovegrove	Mrs R.E. Lovegrove
1973	Miss M.D. Bindley	Mr R.W. Sawyer	Mrs R.E. Lovegrove	Mrs R.E. Lovegrove
1974	Miss M.D. Bindley	Mr R.W. Sawyer	Mrs R.E. Lovegrove	Mrs R.E. Lovegrove

Year				
1975	Miss M.D. Bindley	Mr R.W. Sawyer	Mr M.L. Tovey	Mr M.L. Tovey
1976	Miss M.D. Bindley	Mr R.W. Sawyer	Mr M.L. Tovey	Mr M.L. Tovey
1977	Miss M.D. Bindley	Mr R.W. Sawyer	Mr C.A. Harris	Mr C.A. Harris
1978	Miss M.D. Bindley	Mr R.W. Sawyer	Mr C.A. Harris	Mr C.A. Harris
1979	Miss M.D. Bindley	Mr A.D. Charles	Mr C.A. Harris	Mr C.A. Harris
1980	Miss M.D. Bindley	Mr A.D. Charles	*Lt.Cdr. J.M. Goodman	*Mr A.J. Hepworth
1981	Mr R. W. Sawyer	Mr K.A.T. Edwards	Mr A. Downes	Mr A.J. Hepworth
1982	Mr R. W. Sawyer	Mr K.A.T. Edwards	Mr A. Downes	Mr A.J. Hepworth
1983	Mr L. Gulliford	Mr K.A.T. Edwards	Mr A. Downes	Mr A.J. Hepworth
1984	Mr L. Gulliford	Mr J.M. Duffin	Mrs E. Farnes	Mr W.J. Farnes
1985	Mr L. Gulliford	Mr J.M. Duffin	Mrs E. Farnes	Mr W.J. Farnes
1986	Mr A.D. Charles	Mr J.M. Duffin	Mrs E. Farnes	Mr W.J. Farnes
1987	Mr A.D. Charles	Mr F.J. Horne	Mrs E. Farnes	*Mr J. M. Duffin
1988	Mr A.D. Charles	Mr F.J. Horne	Mrs M. Pusey	Mr J.M. Duffin
1989	Mr C.H.G. Langford	Mr F.J. Horne	Mrs M. Pusey	Mr J.M. Duffin
1990	Mr C.H.G. Langford	Mr J.C. Newcombe	Mrs B.M. Quartly	Mr R.P. Dell
1991	Mr C.H.G. Langford	Mr J.C. Newcombe	Mrs B.M. Quartly	Mr R.P. Dell
1992	Mr A.D. Charles	Mr J.C. Newcombe	Mrs B.M. Quartly	Mr R.P. Dell
1993	Mr A.D. Charles	Col. A.C. Elcomb	Mrs B.M. Quartly	Mr R.P. Dell
1994	Mr A.D. Charles	Col. A.C. Elcomb	Mrs B.M. Quartly	Mr R.P. Dell
1995	Mr J.C. Newcombe	Col. A.C. Elcomb	Mrs B.M. Quartly	Mr R.P. Dell
1996	Mr J.C. Newcombe	Mr N.B. Trood	Mrs B.M. Quartly	Mrs S. Blake
1997	Mr J.C. Newcombe	Mr N.B. Trood	Mrs B.M. Quartly	Mrs S. Blake
1998	Col. A.C. Elcomb	Mr N.B. Trood	*Mrs S. Perkins	Mrs S. Blake
1999	Col. A.C. Elcomb	Mr G. Fisher	Mrs S. Perkins	Mrs S. Blake
2000	Col. A.C. Elcomb	Mr G. Fisher	Mrs S. Perkins	Mrs S. Blake
2001	Mr N.B. Trood	Mr G. Fisher	Mrs S. Perkins	Mrs S. Blake
2002	Mr N.B. Trood	Mr M.T. Blake	Mrs S. Perkins	Mrs S. Blake
2003	Mr N.B. Trood	Mr M.T. Blake	Mrs S. Perkins	Mrs S. Blake
2004	Mrs M.E. Barnes-Gorell	Mr M.T. Blake	Mrs S. Perkins	Mrs S. Blake
2005	Mrs M.E. Barnes-Gorell	Mr A.M. Morrice	Mrs S. Perkins	Mrs S. Blake

* *Acting, pro tem.*

Appendix II: Isle of Wight Disease

Extracts from Minutes of the former Street and Glastonbury Branch

AGM, 12th April 1913

The Secretary's and Visiting Expert's reports were read for the year 1912. Isle of Wight disease was reported in Somerton in June and by October twenty-five stocks, nearly all the bees in that neighbourhood, had died out. The Chairman proposed a resolution which was unanimously carried expressing the urgent need for carrying into law the Bee Diseases Bill, the resolution to be forwarded to the Secretary to the Board of Agriculture.

AGM, 5th June 1915

The season was about average, but owing to I.O.W. disease not as much honey as usual was gathered. There were 104 stocks in Street in the spring of 1914. About half died out during the summer, and fifty more died during the winter, many leaving honey in the hives. Only four stocks remained alive in the spring of 1915. A resolution was passed that all empty hives be disinfected and that the committee be authorised to carry this out. After the business a demonstration with a diseased stock took place, an attempt being made to separate the young bees from the old in the hope that young bees might keep clear of the disease.

AGM, 12th July 1916

The Secretary's report for 1915 was read recording the finest honey season ever known, but very few bees in the district to take advantage of it, 104 stocks in Street having died out in two years, not a single stock being alive at the end of 1915. Seven stray swarms came to Street and Compton: all took the disease but one, a May swarm, which gathered 70lb surplus honey. After the meeting a stock of Italian bees was inspected which had been imported from a neighbouring district where they had lived through the I.O.W. disease without apparently becoming infected.

AGM, 13th July 1917

About 75% of the bees in Somerset have died out in the last three years and in some villages there is not a bee left. The problem is to replace them with healthy bees. Two or three stocks have been brought from Kent for this purpose. They have the character of being able to resist the disease to a certain extent. These bees are to be increased as fast as possible and supplied to members at a reasonable price. Two members have been successful in finding an immune bee, and one states that his bees would survive after being placed on infected combs. This is a great advance as it has been authoritatively stated that a bee never recovers after developing the disease.

AGM, 18th August 1919

Good work has been done by two or three of our members in cultivating a strain of bees capable of resisting the disease. Some have re-started and their bees have again died out, but others have been more successful and there are now in the district seventeen beekeepers with bees in their hives and about seventy would-be beekeepers with empty hives. Three members succeeded in obtaining nucleus stocks through the county restocking scheme and others have them ordered but have not yet been supplied.

AGM, 20th September 1920

The honey season has been the worst we have known. As usual, May was the best honey month. All honey gathered after June 6th was eaten by the bees during the poor weather which followed. Fortunately, we have been allowed 14lb of sugar by the Board of Agriculture for each stock. I.O.W. disease is still with us. Five remaining stocks of the English bees in different parts of the district were all affected in the spring, and by this time are probably all dead. In two cases the disease showing in Italian-English crosses but they have recovered. The pure Italians have not shown any sign of the disease. A good deal of experimenting has been done by beekeepers in Street to try and find out how the disease is spread. We have been endeavouring to prove that the main cause of the spread of the disease is owing to queens mating with drones from diseased hives, the queen carrying back the infection to her own hive. In six cases where there was evidence of the queen mating with a diseased drone the stock began to show signs of the disease a month after mating. The theory is very difficult to prove owning to mating taking place high in the air. We recommend destroying a stock as soon as the symptoms are apparent. This clears out the disease and the drones which might spread it to other stocks.

AGM, 15th August 1921

Mr Bigg-Wither of Wells gave us an interesting lecture on the small mite, *Tarsonemus woodii*, the discovery of which, by Dr. Rennie of Aberdeen, startled the beekeeping world last year. The photo-micrographs shown of the mite were splendid in detail and a very interesting half-hour was spent in discussing its life history and possible ways of getting rid of it. We are very pleased to have the opportunity of obtaining information showing exactly what the I.O.W. disease really is. Votes of thanks concluded one of our most interesting and enjoyable meetings.

Sources of Reference

Anderson, Rev. C. G., *Beekeepers' Alphabet, The,* 1886.

Bantock, Anton, *Last Smyths of Ashton Court, The*, Part 3, the Malago Society.

Bee Craft, Ltd., *Bee Craft*, various volumes of magazine.

Bristol, Somersetshire and South Gloucestershire BKA, *Minutes and Annual Reports.*

British Bee Publications, *British Bee Journal*, various volumes of magazine from 1875.

British Beekeepers' Association, *Council Minute, Item 11,* 28.01.1950.

Brown, R. H., OBE, *Great Masters of Beekeeping*, Bee Books New and Old, 1994.

Cowan, Thos. W., *British Bee-keepers' Association Jubilee*, British Beekeepers' Association, 1927.

Devon Beekeepers' Association, *Beekeeping*, various volumes of magazine.

Hamblin, A.W. T., A.M.I., Mun.E., *Return to the Ice Age*, Summary, 1963.

Herrod-Hempsall, W., F.R.E.S., *Bee-Keeping New and Old,* Volumes 1 and 2, British Bee Journal, 1930 and 1937.

Hopkinson, G.W., *One Hundred Years of Beekeeping Examinations*, Lecture notes, 1982.

Jackson, T. E., B.Sc., *Avon Story, The,* National Honey Show, 1976.

Killick, Dr C. R., *Bee World,* contributions in volumes 5–8 inc., Apis Club, 1923–1927.

Mace, Herbert, F.L.S., *Beekeepers' Associations – a critical Survey,* The Beekeeping Annual Office, Harlow, Essex, 1928.

Morland, John, *Four Pages on Bees*, Village Album Volume 35, No. 4, C and J Clark, 1883.

Showler, K., *Development of Beekeeping Associations in England, Studies 1 and 2, The,* IBRA, Bee World, volumes 76(4), 1995 and 77(1), 1996.

Snelgrove, L.E., *Contributions of the BBKA to the Progress of Bee-keeping*, BBKA, 1956.

Somerset Archaeological and Natural History Society, *Proceedings:* vol. 72, 1926, Obituary for T. W. Cowan, vol. 79, 1933, Obituary for C. Tite.

Somerset Beekeepers' Association, *Minutes and Annual Reports,* 1906–2005.

South-West Counties Joint Consultative Council, *Minutes,* 1973.

Virgil, *Eclogues, Georgics and Aeneid of Virgil, The,* translated by C. Day-Lewis, Oxford University Press, 1966.

Index

A
Abbott, Mr C. N. 7
Adam, Brother 38, 60, 70–72
Addision, Mr R. 144
Anderson, Rev. C. G. 10, 12–14, 16, 17
Appleton, Mr H. M. 18
Ashwin, Mr W. H. 40–42

B
Bache, Mr R. A. 150, 151
Baker, Mr A. R. 4, 150
Ball, Mr R. 94
Barnes-Gorell, Mrs M. E. 6, 79, 84, 89, 123, 124, 154
Bates, Mr D. 86, 90
Batty, Mr K. T. 48, 52, 153
Beck, Mr R. 36, 41, 46, 47, 106, 143
Bell, Mr G. M. 149
Betts, Mr E. J. F. 51, 52, 56, 153
Bigg-Wither, Mr L. 26, 27, 32, 34–36, 39, 47, 105, 127, 137, 138, 152, 156
Bindley, Miss M. D. 52, 53, 66, 73, 109, 110, 118, 153, 154
Blake, Mrs S. 98, 147, 154
Blake, Mr M. T. 154
Bradbury, Mr S. A. 36, 56, 106
Brake, Mr C. H. 67–69
Brewer, Mr J. W. 23, 25
Bromley, Mrs J. 78, 150
Bromley, Mr A. 75
Brown, Col. E. C. 62
Brown, Mr J. 19, 20, 21, 23, 25
Buckley, Mr F. 53
Burchill, Mr S. 146
Burkitt, Rev. W. E. 10, 12
Burton, Mr J. H. 25
Butler, Mr J. B. 18, 20
Butter, Mrs C. M. 88, 92, 93, 96

C
Charles, Mr A. D. 6, 61, 63, 66, 67, 71, 73, 79, 86, 93, 105, 117, 147, 150, 154
Clark, Mr F. J. 11–13
Clark, Mr R. 78
Colthurst, Mrs M. 50, 53, 132, 149
Connor, Mr A. B. 133
Cowan, Mr T. W. 20, 26, 27, 33, 35, 36, 40, 41, 99–103, 105, 107, 126, 152
Crane, Dr E. 110
Currell, Mrs E. 147

D
Davis, Dr I. 57
Dawe, Miss H. 22, 100
Dell, Mr R. P. 81, 83, 84, 147, 154
Dixon, Mr L. 84
Downes, Mr A. 73, 154
Duffin, Mrs E. 77
Duffin, Mr J. M. 75, 77, 83, 154

E
Edwards, Mr K. A. T. 72–75, 91, 93, 95, 147, 154
Elcomb, Col. A. C. 121, 123, 147, 154

F
Farnes, Mrs E. 75, 77, 154
Farnes, Mr W. J. 75, 77, 154
Fieldhouse, Mr J. 76, 150
Fisher, Mr G. 85, 86, 89, 92, 94, 147, 150, 151, 154
Fisher, Mrs S. 95
Frisch, Professor Karl von 110

G
Gammon, Mr S. J. 147
Goodman, Lt. Cdr. J. M. 154
Goolden, Mr P. 150
Graham, Cdr. R. E. 36, 39, 152

Index

Griffin, Mr W. N. — 8, 13
Grist, Mr H. J. — 26, 34, 127
Gulliford, Mrs M. — 115, 116
Gulliford, Mr L. — 53, 70, 114–116, 146, 154

H

Hannam, Mr J. — 62, 146, 147
Harris, Mr C. — 34, 46, 55, 57, 65
Harris, Mr C. A. — 69, 73, 154
Harris, Mr T. J. — 89, 91, 93, 94
Hawken, Mr N. — 67
Hawkins, Mr E. G. — 37, 38, 43, 115
Hender, Mr L. — 52, 56, 57, 62, 65, 107, 108, 122, 153
Hender, Mrs L. — 52, 54, 56, 57, 70, 122
Hepworth, Mr A. J. — 67, 75, 154
Herrod, Mr W. — 20, 23
Herrod-Hempsall, Mr W. — 38, 41, 43, 46, 50, 109, 137, 146
Hill-Dawe, Miss — 21
Hook, Rev. W. — 9, 10
Horgan, Mr S. — 151
Horne, Mr F. J. — 75, 78, 79, 154
Husband, Mr D. A. — 64, 74, 77, 79

J

Jackson, Mr T. — 66, 68
Jenner, Mr G. H. — 53, 146
Jolly, Lt. Col. H. F. — 25, 33, 36, 107, 126, 152
Jones, Mr S. N. — 83, 147
Jordan, Mr S. — 22, 23, 25, 36

K

Killick, Dr C. R. — 37, 39–42, 135
Kilner, Mr B. — 81

L

Langford, Mr C. H. G. — 78, 119, 120, 154
Lewis, Mr E. — 47, 146
Litman, Mr R. — 32, 34, 37
Loud, Mr D. — 67, 68
Lovegrove, Mrs R. E. — 69, 153

M

Mace, Mr H. — 42, 76
Martin, Mr J. — 18, 20
Meade, Mr H. H. — 138–140
Meyer, Mr O. — 66
Milton, Mr W. W. M. — 86, 92, 147, 150
Moore, Mr F. W. — 39
Moore, Mr H. J. — 31
Moore, Mr K. A. J. — 31
Morrice, Mr A. — 89, 93
Morris, Mr D. G. — 63, 65, 78, 146, 151

N

Newcombe, Mr J. C. — 53, 81, 120, 146, 147, 154

P

Pardoe, Mr L. N. — 47, 48, 153
Peel, Rev. H. R. — 10, 103
Perkins, Mrs S. — 147, 154
Poole, Mr O. — 8
Priscott, Mr M. — 74, 75
Pusey, Mrs M. — 77, 154

Q

Quartly, Mrs B.M. — 77, 154

R

Reynolds, Mr J. C. — 62, 153
Rich, Mr P. J. — 6, 119, 147
Rolt, Mr A. C. — 50, 51, 53, 55, 56, 58, 64–67, 69
Rudman, Mr H. J. — 34, 39, 46, 143

S

Sawyer, Mr R. W. — 53, 55, 62, 64, 66–68, 71–73, 75–76, 111, 112, 153, 154
Shore, Major J. Lintorn — 52, 153
Smyth, Lady E. — 18, 27, 99, 100, 152
Snelgrove, Mr L. E. — 23–27, 33–35, 39–42, 44–47, 55, 62, 99, 101, 105–107, 133, 143, 152, 153
Sparks, Mr F. G. — 48, 56, 63
Sparks, Mr T. — 63, 65
Spiller, Mr J. — 26, 39, 41, 54, 56
Stoker, Capt. T. G. — 48, 52, 149, 153

Street, Dr M. 78
Strudwick, Miss F. 105

T

Tite, Mr C. 7–9, 16, 17, 26, 39, 125
Tonsley, Mr C. C. 107
Tovey, Mr M. L. 69, 154
Trafford, Father Aidan 38
Trood, Mr N.B. 81, 85, 86, 122, 123, 147, 154
Trood, Mr T. B. 85, 147
Trump, Mr A. E. 37, 43
Turner, Mr A. D. 42, 43
Tustin, Dr E. M. 62

U

Underwood, Mr F. R. 63

W

Walker, Mr E. I. 32
Walker, Mr E. 41
Wallace, Dr J. 34, 40, 52, 152, 153
Wallis, Master J. 95
Watts, Mrs V. 90
Wawman, Mr S. 150
West, Mr W. 39, 46, 47, 106, 134, 152, 153
Withycombe, Mr W. A. 16, 20–23, 25, 32, 37, 39, 47, 54, 128–130, 132–134, 137, 140, 143, 152
Woodbury, Mr T. W. 101
Wright, Mr D. 59

Y

Yea, Mr P. 57